MATHEMATICAL PROGRAMMING

Steven Vajda

DOVER PUBLICATIONS, INC.
Mineola, New York

Bibliographical Note

This Dover edition, first published in 2009, is an unabridged republication of the work originally published in 1961 by Addison-Wesley Publishing Company, Inc., Reading, Massachusetts.

Library of Congress Cataloging-in-Publication Data

Vajda, S.
    Mathematical programming / Steven Vajda. — Dover ed.
      p. cm.
    Originally published: Reading Mass. : Addison-Wesley, 1961.
    Includes bibliographical references and index.
    ISBN-13: 978-0-486-47213-3
    ISBN-10: 0-486-47213-2
     1. Programming (Mathematics) I. Title.

QA402.5.V32 2009
519.7—dc22

                             2008050085

Manufactured in the United States by Courier Corporation
47213201
www.doverpublications.com

PREFACE

This work has two aims: to provide a textbook of Linear and Non-linear Programming, and to guide the reader to the rapidly expanding frontier of this recent branch of mathematics. In the earlier chapters I have tried to be exhaustive, but always subject to the rule that only fairly elementary mathematics should be required. Later on, some selection of material had to be made conforming to the second aim. It is hoped that the reader will be able to supplement his knowledge by watching the flood of articles on our topics which appear in so many periodicals.

The theoretical foundations are laid in Chapters 2 and 3 (algebra), Chapter 4 (graphs), Chapter 12 (nonlinear programming), and Chapter 13 (dynamic programming). Chapters 5 and 6 are concerned with algorithms, general and special. Chapters 7 and 8 deal with applications, and the remaining chapters with more specific topics, of which Discrete Programming (Chapter 10) is of greatest importance and no doubt remains to be developed much further. The same further development may be expected of Nonlinear Programming, whose presentation in Chapter 12 may well be out of date when this volume appears.

Those who wish to use this volume merely as a textbook of algorithms will make the most use of Chapters 5 and 6. Those to whom references to matrix theory are not quite clear might read the Appendix, which contains all the material required in this context. I have refrained from using the shorthand of matrix notation in order to oblige readers without higher mathematical training.

I have had the great pleasure of discussions with a majority of those authors mentioned in the bibliography, and with others too. I cannot list them all, but it would offend my own sense of fairness and gratitude if I failed to mention my friends Martin Beale, George Dantzig, and Alan Hoffman, who have generously allowed me to call on their time and knowledge and on their brilliant insight, to my great benefit.

An author must accept responsibility for his choice of emphasis. His own interests may have changed during the time when he was struggling with his material. It is hoped that the author's selection, even if found biassed, will not be judged as malevolent. He will be grateful for criticism, however adverse.

Finally, it must be said that this book is not a manual of linear programming computations. This aspect merits a book of its own, and I am not competent to write it. There are many who have the experience and skill required to write it, and I hope they will.

S.V.

CONTENTS

CHAPTER 1

BACKGROUND, HISTORY, AND EXAMPLES

This book deals with mathematical methods applied to problems of planning. They form part of methods used in, and stimulated by, Operational Research, and their development belongs to the last two decades. Our first concern will be with Linear Programming and with the various modifications of it resulting in the large and still rapidly developing field of nonlinear programming. Our presentation will be mathematical. Those who do not wish to be burdened with detail will easily distinguish the landmarks which contain the results of the arguments. For rigorous justification the mathematical shorthand of formulae remains the best tool.

In 1947 George B. Dantzig carried out an analysis of military programming in cooperation with Marshall Wood, and proposed that the interrelationships between the activities of a large organization should be viewed as a type of model of Linear Programming in which the best programme is determined by minimizing a linear function. It was soon recognized that such models can be used in many contexts, and this brought about the great interest in the theoretical as well as the computational aspects of this field.

During June 20 to 24, 1949, a conference was held in Chicago by the Cowles Commission for Research in Economics. At this conference, scientists classifiable as economists, mathematicians, statisticians, administrators, or combinations thereof, pooled their knowledge, experience, and points of view to discuss the theory and practice of efficient utilization of resources. Most of the conference papers were published in 1951 [10].* In an Introduction the editor, Tjalling C. Koopmans, mentions three sources of the ideas leading to the views expressed at the conference: Austrian and German economists concerned with equations of mathematical economics (Schlesinger, Wald, von Neumann), welfare economics (Bergson, Pareto, Hotelling, and others), and work on inter-industry relationship (Leontief).
⁌ The volume contains a number of fundamental papers, of which [38, 39, 40, 72, 95] are particularly relevant to the topics of the present book. The second and third of these papers, by G. B. Dantzig, may without exaggeration be said to mark the beginning of the calculus of

---

* The numbers in brackets refer to the Bibliography at the end of this book.

1

linear programming. (It will be understood that the selection of the papers here mentioned does not imply any judgement on the merits of other papers in the volume.)

A Symposium on Linear Inequalities and Programming was held on June 14 to 16, 1951, in Washington, D.C., sponsored by the U.S. Department of the Air Force and the National Bureau of Standards; yet another, the Second Symposium on Linear Programming was held in the same city and under the same sponsorship on January 27 to 29, 1955. The papers read at this second symposium were published, edited by H. A. Antosiewitz [1]. More recently, a conference on mathematical planning was held at The RAND Corporation of Santa Monica, California, on March 16 to 20, 1959 [17].

The mathematical theory of linear programming is concerned with finding amongst all solutions of a system of linear algebraic inequalities that solution or those solutions which maximize (or minimize) a given linear expression of the same variables. In practice, planning becomes necessary whenever scarce resources are to be allocated in some optimal way. The term "resources" should here be taken in a very general sense, and it depends, of course, on the specific problem which one considers to be the criterion for one way being better than another. In order to give the reader some idea of the scope of linear programming, we mention now a few simple examples which can be formulated as problems of minimization, or maximization, of a linear expression, subject to linear inequalities.

One of the earliest problems belonging within our field is the so-called Transportation Problem which, in its simplest form, may be described as follows [85, 94]:

There are a_i ships in ports P_i, and b_j are required in ports Q_j. It takes c_{ij} days to sail from P_i to Q_j. Assuming that all ships serve the same purpose equally well, how should the sailing routes of the individual ships be arranged so as to make the total of days sailed as small as possible?

Denoting the number of ships to sail from P_i to Q_j by x_{ij}, we note first that they must be integers and must not be negative. Also, we send a_i ships away from P_i, so that

$$\sum_j x_{ij} = a_i \qquad \text{for all } i.$$

We send b_j ships to Q_j, so that

$$\sum_i x_{ij} = b, \qquad \text{for all } j.$$

We must have $\sum_i a_i = \sum_j b_j$, and thus one of these equations is redundant. These are then the conditions to which the x_{ij} are subject. Out of all those which satisfy the conditions, we want to find those which minimize the total time on voyage, i.e.

$$\sum_i \sum_j c_{ij} x_{ij}.$$

Many other problems, not concerned with ships, or even with transport, can be formulated in the same way, for instance, supplies to customers from factories. In such a case the answers need not be expressed as integers.

Another simple example is the following: A firm can produce three types of cloth, say A, B, and C. Two kinds of wool are required for it, say red wool and green wool. One unit length of Type A cloth needs 3 yards of red wool and 1 yard of green wool; one unit length of Type B needs 4 yards of red wool and 3 yards of green wool; and one unit length of Type C needs 1 yard of red wool and 2 yards of green wool. Altogether we have 20 yards of red wool and 10 yards of green wool in our store, and we assume that the income obtained from one unit length of Type A is 3, of Type B is 6, and of Type C is 2. We want to know how the available material should be used, so as to maximize the income from the finished cloth.

Let the lengths of the three types of cloth produced be denoted by x_1, x_2, x_3, referring to A, B, and C respectively. The income is then $3x_1 + 6x_2 + 2x_3$, and the variables are subject to the following constraints:

$$3x_1 + 4x_2 + x_3 \leqslant 20,$$
$$x_1 + 3x_2 + 2x_3 \leqslant 10.$$

These inequalities express the fact that the production is limited by the availabilities of the two types of wool.

CHAPTER 2

THE ALGEBRA OF LINEAR INEQUALITIES

2–1 Definitions. Linear Programming is concerned with maximizing, or minimizing, a linear expression (called the *objective function*) subject to linear *constraints*. The latter may be inequalities or equations. Usually it will be required that the variables have non-negative values.

Consider the system

$$\sum_{i=1}^{n} a_{ij}x \leqslant b_j \qquad (j = 1, \ldots, m),$$

$$x_i \geqslant 0 \qquad (i = 1, \ldots, n).$$

This system of inequalities is satisfied by the same values of the variables as is the system

$$\sum_{i=1}^{n} a_{ij}x_i + x_{n+j} = b_j \qquad (j = 1, \ldots, m),$$

$$x_i \geqslant 0 \qquad (i = 1, \ldots, n, n + 1, \ldots, n + m).$$

We call the x_{n+j}, which have been introduced to convert the first j inequalities into equations, *additional* variables.* If some or all of the inequality signs in the first j inequalities had been the reverse to the above, then additional variables of non-negative values would have had to be subtracted rather than added. Including the additional variables, there are $n + m = N$, say, non-negative variables altogether.

Given a system of m equations in N variables which can be solved for m variables, and given the requirement that all N variables be non-negative, it is possible to construct an equivalent system consisting entirely of inequalities. Let the equations be

$$\sum_{i=1}^{N} a_{ij}x_i = b_j \qquad (j = 1, \ldots, m).$$

If it is possible to solve this system for m variables, say for $x_1, \ldots, x_m$, so that

$$x_i = z_{i0} - \sum_{k=m+1}^{N} z_{ik}x_k \qquad (i = 1, \ldots, m),$$

---

* They are also called *slack* variables, and the original variables *main* variables.

where z_{i0} and the z_{ik} are constants, then it follows that

$$\sum_{k=m+1}^{N} z_{ik}x_k \leqslant z_{i0},$$

because x_i must be non-negative. These inequalities, together with $x_k \geqslant 0$ for $k = m + 1, \ldots, N$, form a set of inequalities equivalent to the original set of constraints.

The number $N = m + n$ is larger than m, but it does not follow that the system always has a solution in non-negative values. For instance, the system

$$x_1 + x_2 + x_3 = 1, \qquad x_1 + x_2 - x_4 = 2, \qquad x_1, x_2, x_3, x_4 \geqslant 0$$

is inconsistent, because $x_1 + x_2$ cannot, at the same time, be smaller than or equal to 1 and larger than or equal to 2.

If a variable is not sign-restricted, then we can replace it by the difference of two sign-restricted variables, say $x = x' - x''$. When an expression, linear in x, is to be optimized, at least one of x' and x'' will become zero. Alternatively, and more simply, we can express x in terms of other variables, using one of the constraints, and eliminate it from the others by substitution. In general, this should not be done with sign-restricted variables, because it would then become difficult to notice when the remaining variables take on values which make the value of an eliminated variable negative.

We consider now a system of non-negative variables, subject to constraints written as equations, using additional variables if necessary. A set of values x_i^0 of the variables x_i $(i = 1, \ldots, N)$ is called a *solution* if it satisfies the equations. If the values are non-negative, then the solution is *feasible*.

It can happen that the solutions form an unbounded set, so that there are solutions with arbitrarily large values of at least one of the variables. For instance, the constraints $x_1 + x_2 - x_3 = 1$ and $x_1 + x_4 = 1$, with non-negative variables, are satisfied by

$$x_2 = 1 - x_1 + x_3 \qquad \text{and} \qquad x_4 = 1 - x_1,$$

where x_1 is any non-negative value not exceeding 1 and x_3 any non-negative value whatsoever, so that the values which x_2 can take are unbounded.

If we set $N - m = n$ variables equal to zero and ignore the terms containing them (we shall call them the excluded variables), then we obtain a set of m equations in m variables. We distinguish the following cases:

(i) The set can be solved for the variables. We have then a *basic* solution, including the variables which we have excluded, and whose values are, of course, zero. If in this solution more than $N - m$ variables have the value zero, then we speak of a *degenerate* solution, and if such solutions exist, then the original set, and every set equivalent to it, is also called degenerate.

(ii) The set cannot be solved uniquely, because the matrix of its coefficients is singular. Then we do not obtain a basic solution, though a solution may exist in which all excluded variables are, in fact, zero. We illustrate this possibility by an example.

EXAMPLE 2–1. Let
$$x_1 + 2x_2 + x_3 + x_4 = 2,$$
$$x_1 + 2x_2 + \tfrac{1}{2}x_3 + x_5 = 2.$$

The values $x_1 = 1$, $x_2 = \tfrac{1}{2}$, $x_3 = x_4 = x_5 = 0$ form a solution, but not a basic one, because the system $x_1 + 2x_2 = 2$ (repeated) has an infinity of solutions.

In a basic solution we call the excluded variables *nonbasic* and the others *basic*. The columns formed of the coefficients of the basic variables are called a *basis*. In a degenerate solution there may be a choice of which m variables should be called basic, but this is not necessarily so. Examples of the former possibility will appear at later stages. An example of the latter case follows.

EXAMPLE 2–2. Let
$$x_1 + x_2 + 2x_3 = 1, \qquad x_1 + x_2 + 3x_3 = 1.$$

Here $x_1 = 1$, $x_2 = x_3 = 0$ is a solution, and x_1 and x_3 are basic. On the other hand, x_1 and x_2 cannot count as such, because the matrix
$$\begin{pmatrix} 1 & 1 \\ 1 & 1 \end{pmatrix}$$
is singular.

We prove now a lemma which will be of use later on. It concerns the question of whether a solution is, under certain circumstances, basic or not.

LEMMA. If we can solve r ($\leqslant m$) of m consistent and independent constraints, written as equations, for r variables, then these values form part of a basic solution, i.e. we can find $m - r$ other variables such that the constraints can be solved for all m of them, after excluding the others.

Proof. Let a $m \times n$ matrix of rank m be given. We denote it by A. Let there be in it a $m \times m$ submatrix of rank r, denoted by B, and denote an $r \times r$ submatrix of B of rank r by C. We want to find a submatrix of A, of rank m, such that it contains C.

To prove the lemma by induction, it is sufficient to show that C must be contained in a nonsingular submatrix of A of order $(r + 1) \times (r + 1)$. Such a submatrix must exist, because a nonsingular $m \times m$ matrix must contain square nonsingular submatrices of all orders less than m.

Let a nonsingular $(r + 1) \times (r + 1)$ submatrix of A be

$$D = \begin{pmatrix} a_{11} & \cdots & a_{r+1,1} \\ \cdot & & \\ \cdot & & \\ \cdot & & \\ a_{1,r+1} & \cdots & a_{r+1,r+1} \end{pmatrix},$$

and let C be (without restriction of generality)

$$C = \begin{pmatrix} b_{11} & \cdots & b_{r1} \\ \cdot & & \\ \cdot & & \\ \cdot & & \\ b_{1r} & \cdots & b_{rr} \end{pmatrix}.$$

Possibly these two matrices overlap.

We shall now prove that adding to C a row

$$b_{1,r+1} \cdots b_{r,r+1}$$

and an appropriate column taken from D produces a nonsingular matrix. In other words, it is impossible that all the following equations should hold:

$$\begin{vmatrix} b_{11} & \cdots & b_{r1} & a_{11} \\ \cdot & & & \\ \cdot & & & \\ \cdot & & & \\ b_{1r} & \cdots & b_{rr} & a_{1r} \\ b_{1,r+1} & \cdots & b_{r,r+1} & a_{1,r+1} \end{vmatrix} = 0,$$

$$\begin{vmatrix} b_{11} & \cdots & b_{r1} & a_{r+1,1} \\ \cdot & & & \\ \cdot & & & \\ \cdot & & & \\ b_{1r} & \cdots & b_{rr} & a_{r+1,r} \\ b_{1,r+1} & \cdots & b_{r,r+1} & a_{r+1,r+1} \end{vmatrix} = 0.$$

Indeed, using Laplace extension in terms of the last columns, we would have

$$B_1 a_{11} + \ldots + B_{r+1} a_{1,r+1} = 0,$$

$$.$$
$$.$$
$$.$$

$$B_1 a_{r+1,1} + \ldots + B_{r+1} a_{r+1,r+1} = 0,$$

where the B's are cofactors and, in particular, $B_{r+1} = C \neq 0$. However, the coefficient matrix of this set of equations is D, and since D is not singular, this set can only have a solution where all B_i are zero. We have reached a contradiction, and thus at least one of the determinants above must be different from zero.

Less formally, we can prove the lemma by saying that having solved r equations for r variables, and having substituted for these in the remaining equations, some further variable must appear in the latter (because the original equations were consistent and independent). We can solve for it using one equation, substitute in the others for it, and proceed thus until all m equations have been solved.

Given any two feasible (not necessarily basic) solutions of the constraints, say $(x_1', \ldots, x_N')$ and $(x_1'', \ldots, x_N'')$, the values $tx_1' + (1-t)x_1''$, $\ldots, tx_N' + (1-t)x_N''$ also form a feasible solution for $0 \leqslant t \leqslant 1$. On the other hand, even if the two given solutions are basic, this linear combination is basic only if t equals either 0 or 1, in which case it coincides with one of the originally given solutions. For nondegenerate solutions this is immediately clear. A linear combination of degenerate solutions might look like a basic one, but this cannot possibly be true because in this case the linear combination would have m or less positive values for any t between 0 and 1, and the solution, when the corresponding $N - m$ variables are excluded, is not unique.

If a system

$$\sum_{i=1}^{N} a_{ij} x_i = b_j \qquad (j = 1, \ldots, m)$$

is degenerate, then by definition we can find m variables, say $x_1, \ldots, x_m$, such that after solving the system for them and setting the remaining variables equal to zero, one of the selected variables will also be zero. Let this be the variable x_m. Then, by Cramer's rule for solving equations by determinants, we have

$$x_m = \begin{vmatrix} a_{11} & \cdots & a_{m-1,1} & b_1 \\ . & & & \\ . & & & \\ . & & & \\ a_{1m} & \cdots & a_{m-1,m} & b_m \end{vmatrix} \div \begin{vmatrix} a_{11} & \cdots & a_{m1} \\ . & & \\ . & & \\ . & & \\ a_{1m} & \cdots & a_{mm} \end{vmatrix} = 0.$$

The denominator cannot be zero, because if it were $x_1, \ldots, x_m$ could not be basic variables. Hence the numerator is zero, and it follows that the right-hand side of the system of equations is linearly dependent on $m - 1$ columns of the left-hand side.

A sufficient condition for degeneracy is the following: Let the set of constraints

$$x_1 + z_{1.m+1}x_{m+1} + \ldots + z_{1N}x_N = x_1^0$$

$$\cdot$$
$$\cdot$$
$$\cdot$$

$$x_m + z_{m.m+1}x_{m+1} + \ldots + z_{mN}x_N = x_m^0$$

be written, and let there be two nonzero coefficients of one of the variables $x_{m+1}, \ldots, x_N$, say $z_{1.m+1}$ and $z_{2.m+1}$, such that $x_1^0/z_{1.m+1} = x_2^0/z_{2.m+1}$; then the system is degenerate. We prove this by showing that it follows that a set with the following $N - m + 1$ variables, viz. $x_1, x_2, x_{m+2}, \ldots, x_N$, equal to zero satisfies the constraints.

If those variables are zero, then we have

$$z_{1.m+1}x_{m+1} = x_1^0,$$

$$z_{2.m+1}x_{m+1} = x_2^0,$$

$$x_3 + z_{3.m+1}x_{m+1} = x_3^0,$$

$$\cdot$$
$$\cdot$$
$$\cdot$$

$$x_m + z_{m.m+1}x_{m+1} = x_m^0.$$

These are m constraints for the $m - 1$ variables $x_3, \ldots, x_m, x_{m+1}$, but there is no contradiction since the first two equations say, in effect, the same thing.

2–2 Basic feasible solutions. We consider now a set of constraints in N non-negative variables and prove the following two theorems.

THEOREM 2–1. If a feasible solution exists, then at least one basic feasible solution also exists.

THEOREM 2–2. Unless the feasible region is unbounded, every feasible solution is a linear combination of basic feasible solutions with non-negative weights, i.e. the variables $x_1, \ldots, x_N$ of any feasible solution can be written

$$x_i = t_1 x_{i1} + \ldots + t_k x_{ik} \qquad (i = 1, \ldots, N),$$

where $t_1 + \ldots + t_k = 1$, $t_1, \ldots, t_k \geqslant 0$ and $(x_{11}, \ldots, x_{N1})$, $\ldots$ $(x_{1k}, \ldots, x_{Nk})$ are all basic feasible solutions.

Proof. Suppose we have a feasible solution $x_1^0, \ldots, x_N^0$ for a system consisting of m equality constraints. If all x_i^0 are zero, then we have already a basic feasible solution. Otherwise we may assume without restricting generality that

$$x_1^0 > 0, \ldots, x_t^0 > 0, \qquad x_{t+1}^0 = 0, \ldots, x_N^0 = 0, \qquad \text{when } t < N,$$

while if $t = N$, then

$$x_1^0 > 0, \ldots, x_N^0 > 0.$$

Consider the matrix

$$\begin{pmatrix} a_{11} & \cdots & a_{t1} \\ \cdot & & \\ \cdot & & \\ \cdot & & \\ a_{1m} & \cdots & a_{tm} \end{pmatrix},$$

in which all rows and t of the columns of the original matrix of coefficients are represented. Let this matrix have rank r. We may assume that the notation is such that the matrix

$$\begin{pmatrix} a_{11} & \cdots & a_{r1} \\ \cdot & & \\ \cdot & & \\ \cdot & & \\ a_{1r} & \cdots & a_{rr} \end{pmatrix}$$

is not singular. We can, therefore, express $x_1, \ldots, x_r$ in terms of the other variables. Thus

$$x_1 = v_1 + w_{1.r+1}x_{r+1} + \ldots + w_{1N}x_N,$$
$$\cdot$$
$$\cdot$$
$$\cdot$$
$$x_r = v_r + w_{r.r+1}x_{r+1} + \ldots + w_{rN}x_N,$$

and this set holds, of course, for $x_1 = x_1^0, \ldots, x_N = x_N^0$.

If $r < m$ and we substitute these values for $x_1, \ldots, x_r$ into the $(r+1)$th, $\ldots$, mth constraint, then we shall find that $x_{r+1}, \ldots, x_t$ will not appear in them anymore. They involve only $x_{t+1}, \ldots, x_N$. Otherwise the rank could not have been r.

If $x_{r+1}^0 = 0$, and hence the variables with higher subscripts are also zero, then we have a set of $r \leqslant m$ positive variables which satisfy the constraints when those remaining are zero; that this is a basic feasible solution follows from the lemma of the previous section. If $x_{r+1}^0 > 0$, then also $x_1^0 > 0, \ldots, x_r^0 > 0$, and since $r + 1 \leqslant t$, the $x_1, \ldots, x_{r+1}$ do not appear in any of the $(r + 1)$th, $\ldots$, mth constraint. They can therefore be ignored if we keep the values of $x_{r+2}, \ldots, x_N$ fixed, and we can slightly decrease x_{r+1} without making any of $x_1, \ldots, x_r$ negative.

We can decrease x_{r+1} to zero or until one of $x_1, \ldots, x_r$ is zero. In either case we have reduced the number of positive x_i and in repeating the process, after renumbering if necessary, we reach eventually a basic feasible solution. This proves Theorem 2-1.

On the other hand, if we increase x_{r+1} and the region of feasible solutions is bounded, then it cannot be increased indefinitely without making some other variable zero. Thus $x_1^0, \ldots, x_N^0$ is seen to be a linear combination of two solutions with less than t positive values of the variables. Continuing, we see that it is a linear combination of basic feasible solutions. This proves Theorem 2-2.

EXAMPLE 2-3. Let

$$3x_1 + 4x_2 + x_3 + x_4 = 2,$$
$$x_1 + 3x_2 + 2x_3 + x_5 = 1.$$

The point $X^0 = (1/6, 1/6, 1/6, 2/3, 0)$ is feasible, but not basic. The rank of

$$\begin{pmatrix} 3 & 4 & 1 & 1 \\ 1 & 3 & 2 & 0 \end{pmatrix}$$

is two, and

$$\begin{pmatrix} 3 & 4 \\ 1 & 3 \end{pmatrix}$$

is not singular. Write

$$3x_1 + 4x_2 = 2 - x_3 - x_4,$$
$$x_1 + 3x_2 = 1 - 2x_3 - x_5,$$

i.e.

$$x_1 = 2/5 + x_0 - 3x_4/5 + 4x_5/5,$$
$$x_2 = 1/5 - x_3 + x_4/5 - 3x_5/5.$$

The term x_3 can be decreased until it is zero, in which case we have $X^1 = (0, 1/3, 0, 2/3, 0)$ and this, as it happens, is already basic. By increasing x_3, we can make it as large as $1/3$ (i.e., as large as $1/5 + x_4/5 - 3x_5/5$ is at present), and we obtain $X^2 = (1/3, 0, 1/3, 2/3, 0)$. We have also $X^0 = \frac{1}{2}(X^1 + X^2)$, but X^2 is not basic itself, and must be further dealt with.

The solution $X^2 = (1/3, 0, 1/3, 2/3, 0)$ is feasible but not basic. The matrix

$$\begin{pmatrix} 3 & 1 \\ 1 & 2 \end{pmatrix}$$

is not singular so we write

$$3x_1 + x_3 \ = 2 - 4x_2 - x_4,$$
$$x_1 + 2x_3 = 1 - 3x_2 - x_5,$$

i.e.

$$x_1 = 3/5 - x_2 - 2x_4/5 + \ x_5/5,$$
$$x_3 = 1/5 - x_2 + \ x_4/5 - 3x_5/5.$$

The term x_4 can be decreased until it is zero, and we have then

$$X_1^2 = (3/5, 0, 1/5, 0, 0);$$

the term x_4 can also be increased, until it is $3/2$, and we have then

$$X_2^2 = (0, 0, 1/2, 3/2, 0),$$

so that

$$X^2 = \frac{5X_1^2 + 4X_2^2}{9}.$$

Finally,

$$X^0 = \frac{X^1}{2} + \frac{5X_1^2}{18} + \frac{2X_2^2}{9},$$

where X^1, X_1^2, and X_2^2 are basic feasible solutions.

Now consider the objective function and the problem of optimizing it. We shall prove two theorems concerning optimizing solutions.

THEOREM 2–3. If a feasible optimizing solution exists, then at least one basic feasible optimizing solution exists as well.

THEOREM 2–4. Unless the feasible region is unbounded, every feasible optimizing solution is a linear combination of basic feasible optimizing solutions.

Proof. We proceed as in the proof of Theorem 2–1, until we reach

$$x_i = v_i + w_{i,r+1}x_{r+1} + \ldots + w_{iN}x_N \qquad (i = 1, \ldots, r).$$

By substitution, we write the objective function B, to be maximized, as

$$B = v_0 + w_{0,r+1}x_{r+1} + \ldots + w_{0N}x_N.$$

If B is to be a maximum, then it must not become larger, however x_{r+1} changes. Hence, if x_{r+1} is positive, then its coefficient $w_{0,r+1}$ must be zero. It follows that we can reach a basic feasible solution, in the way explained above, without changing the (already optimal) value of B. If $x_{r+1} = 0$, then we obtain an optimal basic feasible solution by reference to the previous lemma. This is a useful result, because it tells us that in looking for optimal solutions, we can restrict our search to basic feasible ones.

EXAMPLE 2–4. It is required to maximize

$$B = 3x_1 + 4x_2 + x_3,$$

subject to

$$3x_1 + 4x_2 + x_3 + x_4 = 2 \quad \text{and} \quad x_1 + 3x_2 + 2x_3 + x_5 = 1.$$

(By virtue of the first constraint the maximum is, clearly, 2.) An optimal solution is, for instance, $(1/2, 1/10, 1/10, 0, 0)$ which is not basic. Expressing x_1 and x_2 in terms of the other variables, as before, we have after substitution into B

$$B = 2 - x_4.$$

In this case $r = 2$ and x_3, which has the positive value $1/10$, has a zero coefficient in the expression of B. (An optimal basic solution is X_1^2; another will be exhibited in Example 2–5.)

The above proof is taken from [14]. Another proof, in [109], is as follows: Let

$$x_1^0 > 0, \ldots, x_t^0 > 0, \qquad x_{t+1}^0 = 0, \ldots, x_N^0 = 0$$

be an optimal solution of the equality constraints, i.e. one that maximizes the objective function $b_1 x_1 + \ldots + b_N x_N$. If the matrix

$$\begin{pmatrix} a_{11} & a_{21} & \cdots & a_{t1} \\ \cdot & & & \\ \cdot & & & \\ \cdot & & & \\ a_{1m} & a_{2m} & \cdots & a_{tm} \end{pmatrix}$$

has rank $r < t$ (this is certainly the case when $t > m$), then we can find constants $k_1, \ldots, k_t$, not all zero, such that

$$A(k_1 a_{11} + \ldots + k_t a_{t1}) = 0,$$
$$\cdot$$
$$\cdot$$
$$A(k_1 a_{1m} + \ldots + k_t a_{tm}) = 0,$$

where A is an arbitrary constant. Then, for sufficiently small A, so that none of the variables has negative value, we have also a solution

$$x_1' \quad = x_1^0 + Ak_1,$$

.

.

.

$$x_t' \quad = x_t^0 + Ak_t,$$

$$x_{t+1}' = x_{t+1}^0,$$

.

.

.

$$x_N' \quad = x_N^0.$$

The value of the objective function is now

$$k_1 x_1' + \ldots + b_N x_N' = b_k x_1^0 + \ldots + b_N x_N^0 + A(b_1 k_1 + \ldots + b_t k_t).$$

Because A may be positive or negative, the right-hand side could be made larger than $b_1 x_1^0 + \ldots + b_N x_N^0$, unless $b_1 k_1 + \ldots + b_t k_t = 0$. But we assumed that $x_1^0, \ldots, x_N^0$ was an optimal solution, so that the latter equation must hold. We can therefore choose A so that at least one more of the x_i^0 becomes zero, without altering thereby the value of the objective function. If necessary, we repeat the procedure until $t \leqslant m$.

EXAMPLE 2–5. We use the same case as before and start from the fact that $(1/2, 1/10, 1/10, 0, 0)$ is a feasible optimal solution, though not a basic one. We must therefore find k_1, k_2, and k_3, not all zero, such that

$$3k_1 + 4k_2 + \quad k_3 = 0,$$
$$k_1 + 3k_2 + 2k_3 = 0.$$

We can take, for instance, $k_1 = 1$, $k_2 = -1$, $k_3 = 1$. Therefore $(1/2 + A, 1/10 - A, 1/10 + A, 0, 0)$ is also optimal, for any A, and taking $A = 1/10$ or $A = -1/10$, we obtain, respectively, $(3/5, 0, 1/5, 0, 0)$ or $(2/5, 1/5, 0, 0, 0)$.

To avoid misunderstandings, we may say here, explicitly, that we do not claim to find in this manner *all* optimal basic feasible solutions.

2–3 Transformations. We shall now investigate in more detail the transformations of a system of m equations in N variables, where

$N > m$. At this stage we ignore the requirement that the variables have non-negative values. Let the equations be

$$a_{1j}x_1 + \ldots + a_{Nj}x_N = b_j \qquad (j = 1, \ldots, m),$$

and let the objective function x_0 be defined by the equation

$$-c_1 x_1 - \ldots - c_N x_N + x_0 = 0.$$

We deal with this system of $m + 1$ equations, taking notice, where convenient, of the fact that the variable x_0 appears only in one single equation.

Denote by M the matrix of coefficients

$$\begin{pmatrix} b_1 & a_{11} & \cdots & a_{N1} & 0 \\ b_2 & a_{12} & \cdots & a_{N2} & 0 \\ \cdot & & & & \\ \cdot & & & & \\ \cdot & & & & \\ b_m & a_{1m} & \cdots & a_{Nm} & 0 \\ 0 & -c_1 & \cdots & -c_N & 1 \end{pmatrix}.$$

We select now m variables, say $x_{u_1}, \ldots, x_{u_m}$, such that the first m equations can be solved for them. This means that the matrix

$$\begin{pmatrix} a_{u_1 1} & \cdots & a_{u_m 1} \\ \cdot & & \\ \cdot & & \\ \cdot & & \\ a_{u_1 m} & \cdots & a_{u_m m} \end{pmatrix}$$

is not singular (its determinant is not zero), and then the matrix

$$\begin{pmatrix} a_{u_1 1} & \cdots & a_{u_m 1} & 0 \\ \cdot & & & \\ \cdot & & & \\ \cdot & & & \\ a_{u_1 m} & \cdots & a_{u_m m} & 0 \\ -c_{u_1} & \cdots & -c_{u_m} & 1 \end{pmatrix} = M_{u_1, \ldots, u_m},$$

say, is not singular either. Therefore its inverse exists. We write it, say,

$$\begin{pmatrix} d_{1u_1} & \cdots & d_{mu_1} & 0 \\ \cdot & & & \\ \cdot & & & \\ \cdot & & & \\ d_{1u_m} & \cdots & d_{mu_m} & 0 \\ d_{10} & \cdots & d_{m0} & 1 \end{pmatrix} = D_{u_1, \ldots, u_m}.$$

Then the following relations hold:

$$a_{s1}d_{1t} + \ldots + a_{sm}d_{mt} = \begin{cases} 1 & \text{if } s = t, \\ 0 & \text{if } s \neq t \end{cases} \quad (s, t = u_1, \ldots, u_m), \qquad (2\text{--}1)$$

$$a_{u_1t}d_{su_1} + \ldots + a_{u_mt}d_{su_m} = \begin{cases} 1 & \text{if } t = s, \\ 0 & \text{if } t \neq s \end{cases} \quad (s, t = 1, \ldots, m), \qquad (2\text{--}2)$$

$$d_{s0} = c_{u_1}d_{su_1} + \ldots + c_{u_m}d_{su_m} \qquad (s = 1, \ldots, m). \qquad (2\text{--}3)$$

It should be understood that although only one of the u_i is shown as a subscript of d_{su_i}, its value depends also on the other u_i in $M_{u_1, \ldots, u_m}$.

We solve now the $m + 1$ equations for $x_{u_1}, \ldots, x_{u_m}$ and x_0, and obtain

$$x_{u_i} = \sum_{t=1}^{m} d_{tu_i}(b_t - \sum_{k=m+1}^{N} a_{u_kt}x_{u_k}) \qquad (i = 1, \ldots, m). \qquad (2\text{--}4)$$

This is seen by multiplying the left-hand side by a_{u_ij}, adding over the u_i, and using (2–2). Similarly, we have

$$x_0 = \sum_{t=1}^{m} d_{t0}(b_t - \sum_{k=m+1}^{N} a_{u_kt}x_{u_k}) + \sum_{k=m+1}^{N} c_{u_k}x_{u_k}$$

$$= c_{u_1} \sum_{t=1}^{m} d_{tu_1}b_t + \ldots + c_{u_m} \sum_{t=1}^{m} d_{tu_m}b_t \qquad (2\text{--}5)$$

$$+ \sum_{k=m+1}^{N} x_{u_k}(c_{u_k} - \sum_{s=1}^{m} c_{u_s} \sum_{t=1}^{m} d_{tu_s}a_{u_kt}).$$

This can also be verified by substituting from (2–4) into the expression for x_0.

These relations will be frequently used, and it is therefore convenient to introduce an abbreviated notation, as follows:

$$\sum_{t=1}^{m} d_{tu_i}b_t = x_{u_i}^0 \qquad (i = 1, \ldots, m), \qquad (2\text{--a})$$

$$\sum_{t=1}^{m} d_{tu_i}a_{u_kt} = z_{u_iu_k} \qquad (i = 1, \ldots, m; \ k = m+1, \ldots, N), \qquad (2\text{--b})$$

$$\sum_{i=1}^{m} c_{u_i}x_{u_i}^0 = z_{00}, \qquad (2\text{--c})$$

$$\sum_{t=1}^{m} c_{u_t}z_{u_tu_k} - c_{u_k} = z_{0u_k} \qquad (k = m+1, \ldots, N). \qquad (2\text{--d})$$

With this notation we rewrite (2–4) and (2–5) to read

$$x_{u_i} + \sum_{k=m+1}^{N} z_{u_i u_k} x_{u_k} = x_{u_i}^0 \qquad (i = 1, \ldots, m), \tag{2–6}$$

$$x_0 + \sum_{k=m+1}^{N} z_{0u_k} x_{u_k} = z_{00}. \tag{2–7}$$

If the values of x_{u_k} for $k = m + 1, \ldots, N$ are zero, then the values of those variables for which we have solved are $x_{u_i} = x_{u_i}^0$, and the value of x_0 is, of course,

$$\sum_{i=1}^{m} c_{u_i} x_{u_i}^0 = z_{00}.$$

We shall also make use of the relation

$$a_{u_1 j} z_{u_1 u_k} + \ldots + a_{u_m j} z_{u_m u_k} = \sum_{i=1}^{m} a_{u_i j} \sum_{t=1}^{m} d_{t u_i} a_{u_k t} = a_{u_k j} \tag{2–8}$$

$$(j = 1, \ldots, m; \quad k = m + 1, \ldots, N) \text{ [by (2–2)]}.$$

It follows in the same way that

$$a_{u_1 j} x_{u_1}^0 + \ldots + a_{u_m j} x_{u_m}^0 = b_j \qquad (j = 1, \ldots, m), \tag{2–9}$$

which is obvious from first principles.

Finally, we mention that the product of the matrices $D_{u_1, \ldots, u_m}$ and M is

$$\begin{pmatrix} x_{u_1}^0 & z_{u_1 1} & \cdots & z_{u_1 N} & 0 \\ \cdot & & & & \\ \cdot & & & & \\ \cdot & & & & \\ x_{u_m}^0 & z_{u_m 1} & \cdots & z_{u_m N} & 0 \\ z_{00} & z_{01} & \cdots & z_{0N} & 1 \end{pmatrix},$$

where the column of the second subscript u_s will have a 1 in the $x_{u_s}^0$ row and 0 everywhere else, i.e. $z_{u_s u_s} = 1$ and $z_{u_s u_t} = 0$ for $s \neq t$ $(s, t = 1, \ldots, m)$.

If the m equations from which we started were

$$\sum_{i=m+1}^{N} a_{ij} x_i + x_j = b_j \qquad (j = 1, \ldots, m),$$

$$-c_{m+1} x_{m+1} - \ldots - c_N x_N + x_0 = 0,$$

then the product of $D_{1.2\ldots,m}$ and M is

$$\begin{pmatrix} b_1 & 1 & 0 & \ldots & 0 & a_{m+1,1} & \ldots & a_{N1} & 0 \\ \cdot & & & & & & & & \\ \cdot & & & & & & & & \\ \cdot & & & & & & & & \\ b_m & 0 & 0 & \ldots & 1 & a_{m+1,m} & \ldots & a_{Nm} & 0 \\ 0 & 0 & 0 & \ldots & 0 & -c_{m+1} & \ldots & -c_N & 1 \end{pmatrix}.$$

Assume now that we take two sets of m variables, say $u_1, \ldots, u_m$ and $v_1, \ldots, v_m$. The two sets may have elements in common, and may even be the same. We show now that the two matrices

$$\begin{pmatrix} z_{u_1 v_1} & \ldots & z_{u_1 v_m} & 0 \\ \cdot & & & \\ \cdot & & & \\ \cdot & & & \\ z_{u_m v_1} & \ldots & z_{u_m v_m} & 0 \\ z_{0 v_1} & \ldots & z_{0 v_m} & 1 \end{pmatrix} \quad \text{and} \quad \begin{pmatrix} z_{v_1 u_1} & \ldots & z_{v_1 u_m} & 0 \\ \cdot & & & \\ \cdot & & & \\ \cdot & & & \\ z_{v_m u_1} & \ldots & z_{v_m u_m} & 0 \\ z_{0 u_1} & \ldots & z_{0 u_m} & 1 \end{pmatrix}$$

are inverse to one another. (Again, it must be understood that a value $z_{u_1 v_1}$, for instance, depends also on all the other u_i in the selection.) To prove the statement, we compute

$$z_{u_s v_1} z_{v_1 u_t} + \ldots + z_{u_s v_m} z_{v_m u_t} = \sum_{i=1}^{m} \sum_{k=1}^{m} d_{k u_s} a_{v_i k} \cdot \sum_{r=1}^{m} d_{r v_i} a_{u_t r}$$

$$= \sum_{k=1}^{m} d_{k u_s} \sum_{r=1}^{m} a_{u_t r} \sum_{i=1}^{m} a_{v_i k} d_{r v_i}.$$

The third sum is unity for $k = r$, and zero otherwise, so that the treble sum reduces to

$$\sum_{k=1}^{m} d_{k u_s} a_{u_t k} = \begin{cases} 1 & \text{if } t = s, \\ 0 & \text{if } t \neq s. \end{cases}$$

A similar relationship holds for the last row of each matrix, and thus our statement is completely proved.

Consider now the system

$$x_i + \sum_{k=m+1}^{N} z_{ik} x_k = x_i^0 \qquad (i = 1, \ldots, m),$$

$$x_0 + \sum_{k=m+1}^{N} z_{0k} x_k = z_{00},$$

and let $(x_1^0, \ldots, x_m^0, 0, \ldots, 0)$ be a basic solution. Two basic solutions are called *adjacent* if there is only one basic variable in each which is not contained in the other. We are interested in knowing how the matrix of the $z_{u_i u_k}$ changes if we move from one basic solution to an adjacent one. Let us assume that, in the system above, we want to make x_{m+1} basic instead of x_m. The equation containing x_m gives

$$x_{m+1} + \frac{x_m}{z_{m,m+1}} + \frac{z_{m,m+2}}{z_{m,m+1}} x_{m+2} + \ldots + \frac{z_{mN}}{z_{m,m+1}} x_N = \frac{x_m^0}{z_{m,m+1}}.$$

We call $z_{m,m+1}$ the pivot of the exchange, and the equation which contains the two variables to be exchanged is called the pivotal equation.

We substitute the new expression for x_{m+1} into the other equations and obtain

$$x_i - \frac{z_{i,m+1}}{z_{m,m+1}} x_m + \left(z_{i,m+2} - \frac{z_{i,m+1} \cdot z_{m,m+2}}{z_{m,m+1}} x_{m+2} \right) + \ldots$$

$$\ldots + \left(z_{iN} - \frac{z_{i,m+1} \cdot z_{mN}}{z_{m,m+1}} \right) x_N = x_i^0 - \frac{z_{i,m+1}}{z_{m,m+1}} x_m^0$$

for $i = 1, 2, \ldots, m-1$, and

$$x_0 - \frac{z_{0,m+1}}{z_{m,m+1}} x_m + \left(z_{0m} - \frac{z_{0,m+1} \cdot z_{m,m+2}}{z_{m,m+1}} \right) x_{m+2} + \ldots$$

$$\ldots + \left(z_{0N} - \frac{z_{0,m+1} \cdot z_{mN}}{z_{m,m+1}} \right) x_N = z_{00} - \frac{z_{0,m+1} x_m^0}{z_{m,m+1}}.$$

2–4 Geometric representation. N real values may be considered to be coordinates of a point in N-dimensional space. If these values are subject to m independent and consistent linear equations, then m variables can be eliminated and the points whose N coordinates satisfy the equations can be identified by $n = N - m$ values; they form an n-dimensional space.

If the coordinates are restricted by a linear inequality, say

$$\sum_{i=1}^{n} a_i x_i \leqslant b,$$

then their dimensionality is not thereby restricted. However, further inequalities may restrict it. For instance, for $n = 3$, the points for which $x_1 \geqslant 0$, $x_2 \geqslant 0$, $x_1 + x_2 + x_3 \geqslant 1$ form a three-dimensional continuum. On the other hand, all points whose coordinates satisfy

$x_1 \geqslant 0$, $x_1 \leqslant 0$, $x_1 + x_2 + x_3 \geqslant 1$ lie in the two-dimensional space defined by $x_1 = 0$. Inequalities can, of course, be inconsistent, as we have already seen.

Points whose coordinates satisfy given linear inequalities are called *feasible*. An inequality such as

$$\sum_i a_i x_i \leqslant b$$

restricts them to those on the straight line with equation $\sum_i a_i x_i = b$ and those on one side of it. To establish the correct side, we take some point outside this line (e.g. $x_1 = \ldots = x_n = 0$, unless $b = 0$), and see whether its coordinates satisfy the inequality. If they do, then the side containing that point belongs to the feasible region. We deal thus with all inequalities, restricting the feasible region further and further.

If we demand that the values x_i be non-negative, then we restrict ourselves thereby to the first orthant (in two dimensions to the first quadrant, in three to the first octant) of the space, including its boundary points, of which the origin is one. If we write a constraint, say $\sum_i a_i x_i \leqslant b$ equivalently as $\sum_i a_i x_i + x' = b$, $x' \geqslant 0$, then the hyperplane $\sum_i a_i x_i = b$ is also defined by $x' = 0$. The boundaries of the first orthant are $x_i = 0$ ($i = 1, \ldots, n$), and each linear portion of the boundary of the feasible region satisfies $x_i = 0$ for some of the $m + n = N$ values of i.

The region thus defined by a set of inequalities is *convex*. This means that all points on a straight line between two points of the region belong to that region. This follows from the fact that a linear combination of two feasible solutions is also a feasible solution, as has been mentioned.

A point whose coordinates form a basic feasible solution is called a *vertex*. This corresponds to the geometric idea of a vertex, because it can also be defined as a point which does not lie half-way between any two other points of the feasible region. This will now be shown.

Let P_0 be a feasible point with coordinates $(x_1, \ldots, x_m, 0, \ldots, 0)$ which form a basic solution. Let P_1 have coordinates $(y_1, \ldots, y_m, y_{m+1}, \ldots, y_N)$ and be also feasible. If a third point, P_2, be such that P_0 lies half-way between P_1 and P_2, then P_2 must have coordinates $(2x_1 - y_1, \ldots, 2x_m - y_m, -y_{m+1}, \ldots, -y_N)$. Since P_1 is a feasible point, and hence $y_{m+1}, \ldots, y_N$ are non-negative, P_2 can only be feasible if these values are zero. Hence P_1 and P_2 are also vertices, and since in all three points considered the same x_i ($i = m + 1, \ldots, N$) are zero, these three vertices coincide.

This argument was carried out in N-dimensional space. It applies also to the n-dimensional space.

EXAMPLE 2–6. Let

$$3x_1 + x_2 \geqslant 3, \quad 4x_1 + 3x_2 \geqslant 6, \quad x_1 + 2x_2 \geqslant 2, \quad x_1 \geqslant 0, \quad x_2 \geqslant 0.$$

To represent this system in the plane, we draw the axes of coordinates, and those straight lines defined by replacing the inequality signs by equal signs (Fig. 2–1). In the illustration those sides of the lines

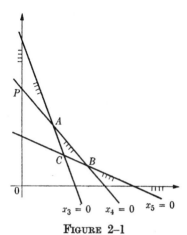

FIGURE 2–1

which are not feasible are marked by short spikes. The additional variables are x_3, x_4, and x_5, in the order in which the inequalities are written. The feasible region is unbounded in this case. It would be bounded if, for instance, we added $x_1 + x_2 \leqslant 6$.

There are here $\binom{5}{2} = 10$ basic solutions, each defined by two of the five variables being zero. They are

the origin:

$$x_1 = x_2 = 0, \quad x_3 = -3, \quad x_4 = -6, \quad x_5 = -2;$$

three points on the vertical axis:

$$x_1 = x_3 = 0, \quad x_2 = 3, \quad x_4 = 3, \quad x_5 = 4;*$$
$$P: \quad x_1 = x_4 = 0, \quad x_2 = 2, \quad x_3 = -1, \quad x_5 = 2;$$
$$x_1 = x_5 = 0, \quad x_2 = 1, \quad x_3 = -2, \quad x_4 = -3;$$

three points on the horizontal axis:

$$x_2 = x_3 = 0, \quad x_1 = 1, \quad x_4 = -2, \quad x_5 = -1;$$
$$x_2 = x_4 = 0, \quad x_1 = 1.5, \quad x_3 = 1.5, \quad x_5 = -0.5;$$
$$x_2 = x_5 = 0, \quad x_1 = 2, \quad x_3 = 3, \quad x_4 = 2;*$$

and, finally, the following three points:

$$A: \quad x_3 = x_4 = 0, \quad x_1 = 3/5, \quad x_2 = 6/5, \quad x_5 = 1;^*$$
$$B: \quad x_4 = x_5 = 0, \quad x_1 = 6/5, \quad x_2 = 2/5, \quad x_3 = 1;^*$$
$$C: \quad x_3 = x_5 = 0, \quad x_1 = 4/5, \quad x_2 = 3/5, \quad x_4 = -1.$$

The four points marked with an asterisk are feasible, and correspond to vertices.

Change now the second constraint to

$$4x_1 + 3x_2 - x_4 = 5.$$

Then the point $x_3 = x_4 = 0$ can also be described as $x_3 = x_5 = 0$, or also as $x_4 = x_5 = 0$ (Fig. 2–2). The system is degenerate.

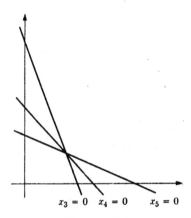

$$x_3 = 0 \quad x_4 = 0 \qquad x_5 = 0$$

FIGURE 2–2

EXAMPLE 2–7. Let

$$x_1 + 2x_2 + x_3 + x_4 = 2,$$
$$x_1 + 2x_2 + \tfrac{1}{2}x_3 + x_5 = 4,$$
$$x_1 + x_2 + x_3 + x_6 = 3.$$

The intersections of $x_4 = 0$ and of $x_5 = 0$ with the plane $x_3 = 0$ are parallel (Fig. 2–3). Hence $x_3 = x_4 = x_5 = 0$ does not define a vertex. The system

$$x_1 + 2x_2 \qquad = 2$$
$$x_1 + 2x_2 \qquad = 4$$
$$x_1 + x_2 + x_6 = 3$$

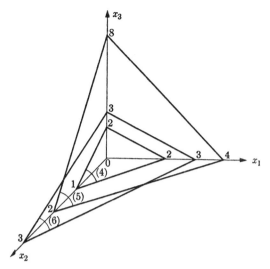

FIGURE 2–3

is, indeed, contradictory. Similarly, the intersections of $x_4 = 0$ and of $x_6 = 0$ with $x_2 = 0$ are parallel, and the system

$$
\begin{aligned}
x_1 + x_3 &= 2 \\
x_1 + \tfrac{1}{2}x_3 + x_5 &= 4 \\
x_1 + x_3 &= 3
\end{aligned}
$$

is also contradictory.

If we introduce new variables by moving the planes $x_1 = 0$, $x_2 = 0$, $x_3 = 0$ parallel to themselves, that is, if

$$
x_1' = x_1 - 6, \quad x_2' = x_2 + 1, \quad x_3' = x_3 + 4,
$$

then we have (omitting the primes)

$$
\begin{aligned}
x_1 + 2x_2 + x_3 + x_4 &= 2, \\
x_1 + 2x_2 + \tfrac{1}{2}x_3 + x_5 &= 2, \\
x_1 + x_2 + x_3 + x_6 &= 2.
\end{aligned}
$$

Now the intersection of $x_4 = 0$ and $x_5 = 0$ falls into the plane $x_3 = 0$, and that of $x_4 = 0$ and $x_6 = 0$ into the plane $x_2 = 0$ (Fig. 2–4). The case is degenerate.

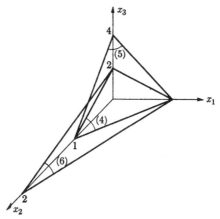

FIGURE 2–4

EXAMPLE 2–8. We have already mentioned the transportation problem. When there are m sources and n destinations, there are $m \cdot n$ variables and $m + n - 1$ independent equations, so that the space in which this problem can be represented is of $mn - m - n + 1 = (m - 1)(n - 1)$ dimensions. As an example, take $m = 2$, $n = 3$. We have then the equations

$$x_{11} + x_{12} + x_{13} = a_1, \qquad x_{11} + x_{21} = b_1,$$
$$x_{21} + x_{22} + x_{23} = a_2, \qquad x_{12} + x_{22} = b_2,$$
$$x_{13} + x_{23} = b_3,$$
$$\text{all } x_{ij} \geqslant 0.$$

Only four of the equations are independent, and they can be written equivalently as follows:

$$x_{21} \leqslant b_1,$$
$$x_{13} \leqslant b_3, \qquad b_1 - a_1 \leqslant x_{21} - x_{13} \leqslant a_2 - b_3.$$

If the constants have the values

$$a_1 = 3, \qquad a_2 = 4, \qquad b_1 = 1, \qquad b_2 = 1, \qquad b_3 = 5,$$

we obtain the diagram of Fig. 2–5.

For other values of a_i and b_j the slanting lines will have other positions; it is interesting to note that if all a_i and b_j are integers, then the coordinates of all vertices are integers too. We have here 12 basic solutions. It can be shown [112] that in a transportation problem with

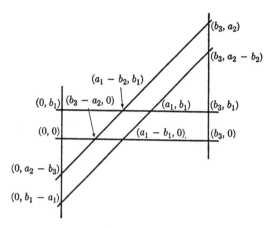

FIGURE 2–5

m sources and n destinations the number of basic solutions is $(m^{n-1})(n^{m-1})$.

Consider now the objective function $C = c_1x_1 + \ldots + c_nx_n$. All those points whose coordinates, substituted into the expression for C, make $C = C_0$, say, define a hyperplane. A change in C_0 induces a parallel displacement of it. All these hyperplanes are perpendicular to the vector from the origin to the point with coordinates $(c_1, \ldots, c_N)$, P_c say. If we want to maximize C, then we call the direction towards P_c *preferred*; if we want to minimize, then the preferred direction is that from P_c towards the origin.

Finding the maximum or the minimum of the objective function means to find that hyperplane, out of a pencil of parallel hyperplanes, which is furthest in the preferred direction and yet has at least one point in common with the feasible region. The coordinates of such a point constitute an *optimal* feasible solution, i.e. one which *optimizes*—maximizes or minimizes, as the case may be—the objective function.

It can happen that an infinity of feasible points lie on the optimal position of the hyperplane. It may also be that there is no finite optimal position. Usually there will be just one point, a vertex of the feasible region, which is optimal.

EXAMPLE 2–9. Let the constraints be as in Example 2–6, with the objective function, to be minimized, $C = 2x_1 + x_2$. The straight line $2x_1 + x_2 = 12/5$ (Fig. 2–6) has exactly one point in common with the feasible region, viz. $A = (3/5, 6/5)$. No value of C is possible lower than $12/5$.

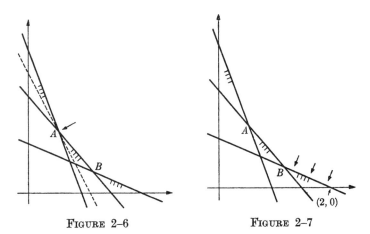

FIGURE 2-6   FIGURE 2-7

If we wanted to maximize C, then any position of a straight line parallel to $2x_1 + x_2 =$ constant could be improved by moving still further out. No bound then exists in the preferred direction. If we wanted to minimize $x_1 + 2x_2$, then all the straight lines corresponding to this objective function are parallel to the boundary $x_1 + 2x_2 = 2$ and all points on this boundary between $B = (6/5, 2/5)$ and $(2, 0)$ produce the minimum, viz. 2 (Fig. 2-7).

EXAMPLE 2-10. Given

$$3x_1 + 4x_2 + x_3 \leqslant 2, \qquad x_1 + 3x_2 + 2x_3 \leqslant 1.$$

$$\text{Maximize } 3x_1 + 6x_2 + 2x_3.$$

This example can be represented in three-dimensional space. The two planes which form the boundary of the feasible region are shown in Fig. 2-8. In the illustration we have also drawn the plane $3x_1 + 6x_2 + 2x_3 = 12/5$, which shows that $12/5$ is the required maximum. It passes through the vertex $(2/5, 1/5, 0)$. (The fact that the optimal value of the objective function is the same as that in Example 2-9 is no coincidence, but it has no importance at this stage. It will be referred to later. See Examples 5-1 and 5-2.)

Another geometric representation arises as follows. A straight line directed from the origin of the system of coordinates to a point with coordinates $(x_1, \ldots, x_n)$ is a vector, and these coordinates can be used, alternatively, to define the vector, or its endpoint. Given m points $(x_{11}, \ldots, x_{n1}) \ldots (x_{1m}, \ldots, x_{nm})$ we say that all points with coordinates

$$(t_1 x_{11} + \ldots + t_m x_{1m}, \ldots, t_1 x_{n1} + \ldots + t_m x_{nm}),$$

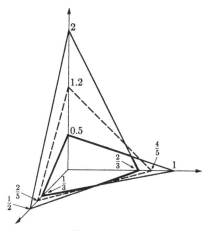

FIGURE 2–8

where $t_1, \ldots, t_m$ are non-negative, form a *polyhedral cone*, spanned by the m vectors with the given coordinates. This cone is convex, has for its vertex the origin, and has for edges some or all the vectors which span it, extended to infinity. Some of the vectors might be inside the cone and could be omitted from the definition without altering the cone.

If we also stipulate that $t_1 + \ldots + t_m = 1$, then we obtain a bounded region of space called the *convex hull* of the given points. In the plane this will be a polygon. Again, some points could be omitted from the definition without altering the region.

It can be shown [62, 126] that a polyhedral cone can equivalently be defined as a region of points whose coordinates satisfy homogeneous linear inequalities. This is not of immediate interest to our present purpose, but we add an example as an illustration.

EXAMPLE 2–11. Let the points (a_{11}, a_{21}) and (a_{12}, a_{22}) be given, and let $a_{11}a_{22} - a_{12}a_{21}$ be positive (Fig. 2–9). Consider in the plane the points with coordinates $(\sum_i a_{1i}y_i, \sum_i a_{2i}y_i)$, where all y_i are non-negative. Consider, also, the points with coordinates (x_1, x_2) satisfying the conditions

$$a_{11}x_2 - a_{21}x_1 \geqslant 0, \qquad -a_{12}x_2 + a_{22}x_1 \geqslant 0.$$

The two regions are identical.

If a vector in n-dimensional space is in the cone spanned by m ($\geqslant n$) vectors, then it can be expressed as a linear combination, with non-negative coefficients, of not more than n of them. If a point in

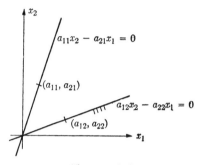

FIGURE 2–9

n-dimensional space is in the convex hull of m ($>n$) points, then it can be expressed as a linear combination with non-negative coefficients, adding up to 1, of not more than $n + 1$ of them. This is so because if a feasible solution exists to a linear programming problem, then a basic feasible solution exists as well (Theorem 2–1). Hence, if

$$x_{11}t_1 + \ldots + x_{1m}t_m = x_1,$$
$$\cdot$$
$$\cdot$$
$$\cdot$$
$$x_{n1}t_1 + \ldots + x_{nm}t_m = x_n,$$
$$t_1, \ldots, t_m \geqslant 0,$$

where the x_{ij} and the x_i are given, has a solution for the t_i, then it also has one with not more than n of the t_i having positive values.

The statement for the convex hulls follows from the fact that if we add the constraint $t_1 + \ldots + t_m = 1$, then a basic feasible solution will have not more than $n + 1$ positive values of its variables.

We can now introduce another geometric interpretation of the linear programming problem. Let it be required to minimize

$$C = c_1x_1 + \ldots + c_Nx_N,$$

subject to

$$\sum_{i=1}^{N} a_{ij}x_i = b_j \qquad (j = 1, \ldots, m),$$
$$x_i \geqslant 0 \qquad (i = 1, \ldots, N).$$

Consider the N points in $(m + 1)$-dimensional space:

$$(a_{11}, a_{12}, \ldots, a_{1m}, c_1) \ldots (a_{N1}, a_{N2}, \ldots, a_{Nm}, c_N).$$

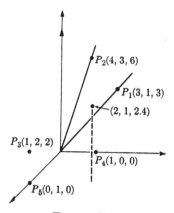

FIGURE 2–10

Because the variables x_i must be non-negative, we concentrate on points of the polyhedral cone defined by the N points. The constraints, which are equations, mean that we consider, amongst the points of that cone, those with coordinates $(b_1, \ldots, b_m, C)$ and in particular that point or those points with the lowest value of C. In a maximizing problem, of course, one would look for the point with the highest value of the last coordinate.

EXAMPLE 2–12. Maximize

$$B = 3x_1 + 6x_2 + 2x_3,$$

subject to

$$3x_1 + 4x_2 + x_3 + x_4 = 2,$$
$$x_1 + 3x_2 + 2x_3 + x_5 = 1.$$

See Fig. 2–10.

EXERCISES

2–1. Find the basic solutions of the following problem:

$$x_1 + x_2 + x_3 \qquad\qquad\qquad = 4,$$
$$x_4 + x_5 + x_6 = 5,$$
$$x_1 \qquad\qquad + x_4 \qquad\quad = 3,$$
$$x_2 \qquad\qquad + x_5 \quad\ = 3.$$

Indicate which of these solutions are feasible.

2-2. Find the basic solutions of

$$x_1 + x_2 + x_3 \quad\;\; = 1,$$
$$3x_1 + 2x_2 \quad\;\; - x_4 = 6.$$

2-3. Change the second equation in Exercise 2-2 to

$$3x_1 + 2x_2 - x_4 + x_5 = 6,$$

and find all basic solutions.

2-4. Find the basic solutions of

(a) $2x_1 + x_2 = 6,$ (b) $2x_1 + x_2 + z_1 = 6,$
     $4x_1 + 2x_2 = 12.$ $4x_1 + 2x_2 + z_2 = 12.$

2-5. Given

$$x_1 + 2x_2 + x_3 + x_4 = 2,$$
$$x_1 + 2x_2 + \tfrac{1}{3}x_3 + x_5 = 2,$$
$$x_1 + x_2 + x_3 + x_6 = 2,$$

with all variables non-negative. Is $(1, \tfrac{1}{2}, 0, 0, 0, \tfrac{1}{2})$ a vertex? If not, express it as a combination of two vertices.

CHAPTER 3

THE ALGEBRA OF DUALITY

In this chapter we deal with the central concept of linear programming algebra: *Duality.*

3–1 Definitions. Two linear programming problems are called dual if they are connected in the following way:

Minimizing problem	Maximizing problem
$a_{11}x_1 + \ldots + a_{n1}x_n \geqslant b_1,$	$a_{11}y_1 + \ldots + a_{1m}y_m \leqslant c_1,$

$$a_{1m}x_1 + \ldots + a_{nm}x_n \geqslant b_m, \qquad a_{n1}y_1 + \ldots + a_{nm}y_m \leqslant c_n,$$
$$\text{all } x_i \geqslant 0. \qquad\qquad \text{all } y_j \geqslant 0.$$

Minimize $c_1x_1 + \ldots + c_nx_n = C.$ Maximize $b_1y_1 + \ldots + b_my_m = B.$

Let us state explicitly the relationship between the two systems:

(i) The objective function of one is to be minimized, that of the other is to be maximized.

(ii) In the minimizing system the inequality sign is $\geqslant$; in the maximizing system it is $\leqslant$, in all constraints.

(iii) The matrix of coefficients of one system of inequalities is the transposed of that of the other system.

(iv) The constants on the right-hand side of one system are the same as the coefficients of the objective function of the other.

It should be noted that the variables are to be non-negative in both systems. We think of duality as a symmetrical relation, and refer to either system as the primal, and then to the other as the dual.

If we have a minimizing system with the inequality signs $\leqslant$, then we can obtain the standard form shown above by changing the signs in the objective function, and maximizing it. The same applies, *mutatis mutandis*, to the dual system. Hence the following two systems are also dual:

$a_{11}x_1 + \ldots + a_{n1}x_n \leqslant b_1,$	$a_{11}y_1 + \ldots + a_{1m}y_m \geqslant -c_1,$

$$a_{1m}x_1 + \ldots + a_{nm}x_n \leqslant b_m, \qquad a_{n1}y_1 + \ldots + a_{nm}y_m \geqslant -c_n,$$
$$\text{all } x_i \geqslant 0. \qquad\qquad \text{all } y_j \geqslant 0.$$

Minimize $c_1x_1 + \ldots + c_nx_n.$ Maximize $-b_1y_1 - \ldots - b_my_m.$

If the primal system also contains equations, e.g.

$$a_{11}x_1 + \ldots + a_{n1}x_n = b_1,$$
$$\cdot$$
$$\cdot$$
$$\cdot$$
$$a_{1k}x_1 + \ldots + a_{nk}x_n = b_k,$$

then we may imagine that each equation is replaced by two inequalities,

$$\left. \begin{array}{c} a_{1j}x_1 + \ldots + a_{nj}x_n \geqslant b_j \\ -a_{1j}x_1 - \ldots - a_{nj}x_n \geqslant -b_j \end{array} \right\} \quad (j = 1, \ldots, k),$$

and the dual system can then be written

$$a_{i1}z_1 - a_{i1}z_1' + \ldots + a_{ik}z_k - a_{ik}z_k' + a_{ik+1}y_{k+1} + \ldots + a_{im}y_m \leqslant c_i$$
$$(i = 1, \ldots, n).$$

The variables z_j, z_j', and y_j must have non-negative values. Because z_j and z_j' have coefficients of different sign but of the same absolute value, we can introduce a new variable,

$$y_j = z_j - z_j' \quad \text{for all} \quad j = 1, \ldots, k,$$

which is not sign restricted, since if y_j is not zero, then either z_j or z_j' may have the larger (non-negative) value. The dual set can therefore be written as before, but only $y_{k+1}, \ldots, y_n$ are to be non-negative, while the signs of those variables $y_1, \ldots, y_k$ referring to the equations in the primal system are not restricted.

The special relationship between two dual systems is reflected in a special relationship between their solutions, and their optimal solutions as well. This will be developed in the course of this chapter.

To begin with, we point to a rather obvious connection. In the minimizing problem, multiply the jth inequality by y_j, and add the resulting inequalities. In the maximizing problem, multiply the ith inequality by x_i, and add the resulting inequalities. The left-hand side will then be

$$\sum_i \sum_j a_{ij}x_iy_j$$

in both cases, and a comparison of the right-hand sides shows that

$$\sum_j b_jy_j \leqslant \sum_i c_ix_i.$$

This inequality holds for any solutions, and hence also for the optimal solutions, of the two systems. We shall prove that for the latter the two sides are in fact equal. Hence optimal solutions are characterized by

$$\sum_j b_jy_j \geqslant \sum_i c_ix_i.$$

EXAMPLE 3–1. Examples 2–9 and 2–10 are dual to one another. The two optimal values of the respective objective functions are equal.

3–2 Homogeneous systems. Our more detailed investigation starts with homogeneous dual systems, i.e. such where the right-hand sides are all zero. To begin with we shall not be concerned with any objective function. Consider, then, the following two dual homogeneous systems :*

$$a_{1j}x_1 + \ldots + a_{nj}x_n \leqslant 0 \qquad\qquad a_{i1}y_1 + \ldots + a_{im}y_m \geqslant 0$$
$$(j = 1, \ldots, m), \qquad\qquad\qquad (i = 1, \ldots, n),$$
$$x_i \geqslant 0 \quad (i = 1, \ldots, n). \qquad\qquad y_j \geqslant 0 \quad (j = 1, \ldots, m).$$

We start by proving a theorem which R. A. Good calls the Key-Theorem, because of its fundamental importance. This proof is based on an unpublished argument due to D. Gale.

THEOREM 3–1. The systems

$$a_{i1}y_1 + \ldots + a_{im}y_m \geqslant 0 \qquad (i = 1, \ldots, n)$$

and

$$a_{1j}x_1 + \ldots + a_{nj}x_n = 0 \qquad (j = 1, \ldots, m)$$
$$x_i \geqslant 0 \qquad\qquad\qquad\qquad (i = 1, \ldots, n)$$

have solutions such that

$$x_i + a_{i1}y_1 + \ldots + a_{im}y_m > 0 \qquad (i = 1, \ldots, n).$$

(Note that the latter relations are *strict* inequalities, and also that the y_j are not sign-restricted.)

Proof. First, we prove the existence of solutions which satisfy the latter strict inequality for $i = 1$. We proceed by induction. The statement is true for $n = 1$. If all a_{1j} are zero, then we may choose $y_j = 0$, $x_1 = 1$; otherwise we choose $y_j = a_{1j}$, $x_1 = 0$.

Let it then be assumed that the statement is proved for a certain value n. Consider now

$$a_{i1}y_1 + \ldots + a_{im}y_m \geqslant 0 \qquad (i = 1, \ldots, n, n + 1),$$
$$a_{1j}x_1 + \ldots + a_{nj}x_n + a_{n+1j}x_{n+1} = 0 \qquad (j = 1, \ldots, m),$$
$$x_1, \ldots, x_n, x_{n+1} \geqslant 0.$$

---

* The development of ideas which follows is mainly based on A. W. Tucker [117] and A. J. Goldman and A. W. Tucker [74]. See also R. A. Good [76].

By our assumption we can find values x_i^0 $(i = 1, \ldots, n)$ and y_j^0 $(j = 1, \ldots, m)$ such that

$$a_{i1}y_1^0 + \ldots + a_{im}y_m^0 \geqslant 0 \qquad (i = 1, \ldots, n),$$
$$a_{1j}x_1^0 + \ldots + a_{nj}x_n^0 = 0 \qquad (j = 1, \ldots, m),$$
$$x_1^0, \ldots, x_n^0 \geqslant 0,$$

and

$$x_1^0 + a_{11}y_1^0 + \ldots + a_{1m}y_m^0 > 0.$$

If

$$a_{n+11}y_1^0 + \ldots + a_{n+1m}y_m^0 \geqslant 0,$$

then the above x_i^0 and y_j^0, together with $x_{n+1}^0 = 0$ satisfy the statement for $n + 1$ as well. On the other hand, if

$$a_{n+11}y_1^0 + \ldots + a_{n+1m}y_m^0 < 0,$$

then these values would not serve. We apply then the statement for the value n, which is assumed to have already been proved, to the systems

$$(a_{i1} + r_i a_{n+11})y_1 + \ldots + (a_{in} + r_i a_{n+1m})y_m \geqslant 0 \qquad (i = 1, \ldots, n),$$
$$(a_{1j} + r_1 a_{n+1j})x_1 + \ldots + (a_{nj} + r_n a_{n+1j})x_n = 0 \qquad (j = 1, \ldots, m),$$
$$x_i \geqslant 0 \qquad (i = 1, \ldots, n),$$

where

$$r_i = \frac{-(a_{i1}y_1^0 + \ldots + a_{im}y_m^0)}{a_{n+11}y_1^0 + \ldots + a_{n+1m}y_m^0} \geqslant 0.$$

This application yields solutions, say $y_j = y_j'$ and $x_i = x_i'$. Then it is seen that

$$y_j = y_j' + ry_j^0, \quad \text{where} \quad r = \frac{-\sum_j a_{n+1j}y_j'}{\sum_j a_{n+1j}y_j^0}$$

together with

$$x_i = x_i' \quad (i = 1, \ldots, n) \qquad \text{and} \qquad x_{n+1} = \sum_i r_i x_i'$$

extend the validity to $n + 1$.

By renumbering the subscripts it follows that there exist solutions $x_{ik}^0 \geqslant 0$, y_{jk}^0 of the systems

$$a_{i1}y_1 + \ldots + a_{im}y_m \geqslant 0 \qquad (i = 1, \ldots, n),$$
$$a_{1j}x_1 + \ldots + a_{nj}x_n = 0 \qquad (j = 1, \ldots, m),$$
$$x_i \geqslant 0 \qquad (i = 1, \ldots, n),$$

such that

$$x_{kk} + a_{k1}y_{1k}^0 + \ldots + a_{km}y_{mk}^0 > 0,$$

and then

$$y_j = \sum_k y_{jk}^0 \quad \text{and} \quad x_i = \sum_k x_{ik}^0$$

satisfy all the conditions of the theorem.

It will be noticed that the proof given is constructive: it tells us how to find the solution. However, it is often simpler to find one by inspection. (See Exercises 3–1 and 3–2.)

THEOREM 3–2. Let $n = n_1 + n_2$, and write the systems of Theorem 3–1 as follows:

$$a_{i1}y_1 + \ldots + a_{im}y_m \geqq 0 \qquad\qquad (i = 1, \ldots, n_1), \qquad \text{(a)}$$
$$a_{i1}y_1 + \ldots + a_{im}y_m \geqq 0 \qquad\qquad (i = n_1 + 1, \ldots, n), \qquad \text{(b)}$$
$$a_{1j}x_1 + \ldots + a_{n_1j}x_{n_1} + \ldots + a_{nj}x_n = 0 \qquad (j = 1, \ldots, m), \qquad \text{(c)}$$
$$x_i \geqq 0 \qquad\qquad (i = 1, \ldots, n).$$

Then (i) either there is a solution for the y_j such that not all relations in (a) are equations, or there is a solution for the x_i satisfying (c) such that all of them are positive for $i = 1, \ldots, n_1$, and (ii) either there is a solution for the y_j such that all relations in (a) are strict inequalities, or there is a solution for the x_i satisfying (c) such that not all x_i are zero. The two alternatives in (i) as well as those in (ii) are mutually exclusive.

Proof. By Theorem 3–1 there exists a solution x_{i1}^0, y_j^0 of the system such that

$$x_i^0 + a_{i1}y_1^0 + \ldots + a_{im}y_m^0 > 0 \qquad (i = 1, \ldots, n_1), \qquad \text{(d)}$$

and also

$$x_i^0 + a_{i1}y_1^0 + \ldots + a_{im}y_m^0 > 0 \qquad (i = n_1 + 1, \ldots, n),$$
$$\text{all } x_i^0 \geqq 0.$$

(i) If for y_j^0 all relations in (a) are equations, then a set $x_i^0 > 0$ exists for $i = 1, \ldots, n_1$; on the other hand if, say,

$$a_{i_01}y_1^0 + \ldots + a_{i_0m}y_m^0 > 0 \qquad (1 \leqq i_0 \leqq n_1),$$

then at least $x_{i_0}^0$ must be zero, since otherwise we would have

$$\sum_{i=1}^n \sum_{j=1}^m x_i^0 a_{ij} y_j^0 > 0,$$

which is impossible, because (c) shows that this double sum must be zero. Hence, also, the two alternatives are mutually exclusive.

(ii) If all $x_i = 0$ for $i = 1, . . ., n_1$ in all solutions of (c), then

$$a_{i1}y_1 + . . . + a_{im}y_m > 0$$

has a solution by (d); if not all $x_i = 0$ for $i = 1, . . ., n_1$, then at least one of the relations (a) must be an equation, for a reason similar to that in (i). H. A. Antosiewitz has shown, in [18], that (i) implies, and is implied by, (ii).

Theorem 3-2 implies a number of special cases which have been known for a long time. If $n_1 = n$, then (i) reduces to a theorem by Stiemke [113]. His paper contains also the special case of (ii) for $n_1 = n$, which was proved by P. Gordan [77].

EXAMPLE 3-2. If $m = 2$, and $n_1 = n$, then the alternatives mentioned in Theorem 3-2 (ii) can be illustrated in the plane (Fig. 3-1). Consider points with coordinates (a_{i1}, a_{i2}), $(i = 1, . . ., n)$.

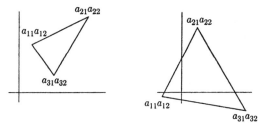

FIGURE 3-1

Either there is a line such that all points lie on the same side of it (and none on it), or the origin can be represented as a linear combination of the points, with non-negative weights throughout.

If some of the inequalities for $i > n_1$ are repeated with reversed inequality sign, so that they hold as equations, and so that hence the corresponding dual variables are not sign-restricted, then (i) reduces to the Transposition Theorem by T. S. Motzkin [105]. The most important specialization for our purpose is that for $n_1 = 1$. It yields the theorem of J. Farkas [62]. Because of the importance of this theorem, we give here its proof as a direct consequence of Theorem 3-1.

FARKAS' THEOREM. If all solutions of the system

$$a_{i1}y_1 + . . . + a_{im}y_m \geqslant 0 \qquad (i = 1, . . ., n)$$

satisfy also the inequality

$$a_{01}y_1 + \ldots + a_{0m}y_m \geqslant 0,$$

then there exist non-negative $t_1, \ldots, t_n$ such that

$$a_{0j} = a_{1j}t_1 + \ldots + a_{nj}t_n \qquad (j = 1, \ldots, m).$$

In geometric terms, we can express this by saying that if all the points which are on those sides of all hyperplanes

$$a_{i1}y_1 + \ldots + a_{im}y_m = 0 \qquad (i = 1, \ldots, n),$$

defined by the vectors $(a_{i1}, \ldots, a_{im})$, lie also on that side of

$$a_{01}y_1 + \ldots + a_{0m}y_m = 0,$$

defined by the vector $(a_{01}, \ldots, a_{0m})$, then the latter vector lies within the cone spanned by the former vectors (Fig. 3–2).

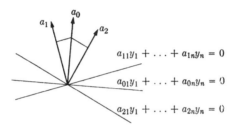

$$a_{11}y_1 + \ldots + a_{1n}y_n = 0$$
$$a_{01}y_1 + \ldots + a_{0n}y_n = 0$$
$$a_{21}y_1 + \ldots + a_{2n}y_n = 0$$

FIGURE 3–2

Proof. Introduce the variable x_0 and state Theorem 3–1 as follows: The system

$$-a_{01}y_1 - \ldots - a_{0m}y_m \geqslant 0$$
$$a_{i1}y_1 + \ldots + a_{im}y_m \geqslant 0 \qquad (i = 1, \ldots, n)$$
$$-a_{0j}x_0 + a_{1j}x_1 + \ldots + a_{nj}x_n = 0 \qquad (j = 1, \ldots, m)$$
$$x_i \geqslant 0 \qquad (i = 0, 1, \ldots, n)$$

has a solution such that

$$x_0 - a_{01}y_1 - \ldots - a_{0m}y_m > 0,$$
$$x_i + a_{i1}y_1 + \ldots + a_{im}y_m > 0 \qquad (i = 1, \ldots, n).$$

In such a solution we have either $x_0 = 0$ or $x_0 > 0$. In the former case it follows that

$$a_{01}y_1 + \ldots + a_{0m}y_m > 0.$$

Therefore if, whenever

$$a_{i1}y_1 + \ldots + a_{im}y_m \geqslant 0 \qquad (i = 1, \ldots, n),$$

we have also

$$a_{01}y_1 + \ldots + a_{0m}y_m \geqslant 0,$$

then the system above which, by Theorem 3–1, still has a solution must have one with $x_0 > 0$. Thus the set of equations above may be divided by x_0 and can be written

$$a_{0j} = a_{1j}t_1 + \ldots + a_{nj}t_n,$$

where the $t_1, \ldots, t_n$ are non-negative.

The inverse of Farkas' theorem is clearly also valid. If there exist non-negative $t_1, \ldots, t_n$ such that

$$a_{0j} = a_{1j}t_1 + \ldots + a_{nj}t_n \qquad (j = 1, \ldots, m),$$

then $a_{01}y_1 + \ldots + a_{0m}y_m \geqslant 0$ is a consequence of

$$a_{i1}y_1 + \ldots + a_{im}y_m \geqslant 0 \qquad (i = 1, \ldots, n).$$

THEOREM 3–3. The systems

$$a_{i1}y_1 + \ldots + a_{im}y_m \geqslant 0 \qquad (i = 1, \ldots, n)$$
$$y_j \geqslant 0 \qquad (j = 1, \ldots, m)$$

and

$$a_{1j}x_1 + \ldots + a_{nj}x_n \leqslant 0 \qquad (j = 1, \ldots, m)$$
$$x_i \geqslant 0 \qquad (i = 1, \ldots, n)$$

have solutions such that

$$y_j - a_{1j}x_1 - \ldots - a_{nj}x_n > 0 \qquad \text{for all } j,$$

and

$$x_i + a_{i1}y_1 + \ldots + a_{im}y_m > 0 \qquad \text{for all } i.$$

This follows from an application of Theorem 3–1, after writing

$$a_{1j}x_1 + \ldots + a_{nj}x_n + w_j = 0, \qquad w_j \geqslant 0 \quad \text{for all } j$$

instead of the equations in x_i.

THEOREM 3–4. Let the matrix

$$\begin{pmatrix} a_{11} & a_{12} & \ldots & a_{1n} \\ \cdot & & & \\ \cdot & & & \\ \cdot & & & \\ a_{n1} & a_{n2} & \ldots & a_{nn} \end{pmatrix}$$

be skew-symmetric, i.e. such that $a_{ij} = -a_{ji}$ for all i, j (and hence, of course, $a_{ii} = 0$). Then Theorem 3–3 states that the systems

$$\left.\begin{array}{l} a_{i1}y_1 + \ldots + a_{in}y_n \geqslant 0 \\ a_{i1}x_1 + \ldots + a_{in}x_n \geqslant 0 \\ x_i \geqslant 0, \qquad\quad y_i \geqslant 0 \end{array}\right\} (i = 1, \ldots, n)$$

have solutions such that

$$x_i + a_{i1}y_1 + \ldots + a_{in}y_n > 0,$$

and

$$y_i + a_{i1}x_1 + \ldots + a_{in}x_n > 0.$$

If we write w_i for $x_i + y_i$, then we can say that the system

$$a_{i1}w_1 + \ldots + a_{in}w_n \geqslant 0 \qquad (i = 1, \ldots, n)$$

has a solution such that

$$w_i + a_{i1}w_1 + \ldots + a_{in}w_n > 0 \qquad (i = 1, \ldots, n).$$

This is the statement of Theorem 3–4.

Of these solutions either $w_i > 0$ and $a_{i1}w_1 + \ldots + a_{in}w_n = 0$, or $w_i = 0$ and $a_{i1}w_1 + \ldots + a_{in}w_n > 0$, for all i. Indeed, at least one of w_i and $a_{i1}w_1 + \ldots + a_{in}w_n$ must be positive, by Theorem 3–4 which has just been proved, and both cannot be, because $\sum_i\sum_j w_i a_{ij} w_j = 0$ since the matrix of the a_{ij} is skew-symmetric.

3–3 Polarity. We think now of the points defined in m-dimensional space as end points of vectors from the origin. Let n points $(a_{i1}, \ldots, a_{im})$ $(i = 1, \ldots, n)$ be given. We are particularly interested in the vectors from the origin to all points with coordinates $(\sum_i a_{i1}x_i, \ldots, \sum_i a_{im}x_i)$, where the x_i are non-negative. The aggregate of all these vectors is a polyhedral convex cone. (See Chapter 2.)

The cone which is *polar* to the latter is defined as consisting of all those vectors from the origin which form nonobtuse angles with all vectors of the polyhedral convex cone. In algebraic terms, the polar cone is defined by those vectors to points with coordinates $y_1, \ldots, y_m$ such that

$$y_1 \sum_{i=1}^{n} a_{i1}x_i + \ldots + y_m \sum_{i=1}^{n} a_{im}x_i \geqslant 0.$$

Denote the original polyhedral cone by (A) and its polar by $(A)^*$. We prove now that $[(A)^*]^*$, i.e. the polar of the polar, which we shall more simply denote by $(A)^{**}$, is again (A).

Clearly, $(A)**$ contains (A), because the polar of $(A)*$ consists of all vectors to points $z_1, \ldots, z_m$ such that

$$z_1y_1 + \ldots + z_my_m \geqslant 0,$$

and

$$z_j = \sum_{i=1}^{n} a_{ij}x_i$$

satisfy this condition.

The converse, i.e. that (A) contains $(A)**$ (and is hence identical with it) follows from Farkas' theorem. The latter states that if for all $y_1, \ldots, y_m$ such that

$$\sum_{j=1}^{m} a_{ij}y_j \geqslant 0 \qquad (i = 1, \ldots, n),$$

or, in other words, such that

$$\sum_i \sum_j a_{ij}y_jx_i \geqslant 0$$

for any set of non-negative x_i we have also $\sum_j c_jy_j \geqslant 0$, then $c_j = \sum_i a_{ij}t_i$ with non-negative t_i. But this means that the vectors defined by $c_1, \ldots, c_m$, which are the vectors defining $(A)**$, belong to (A). We remark that the first part of the proof holds if we start from any set of vectors, not necessarily from a set forming a polyhedral convex cone, but the applicability of Farkas' theorem depends on properties of the latter.

EXAMPLE 3–3. In Fig. 3–3 $n = 2$, $m = 4$.

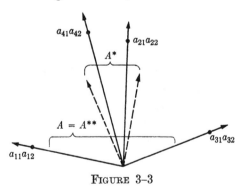

FIGURE 3–3

Relations of the type mentioned above are contained in [70].

3–4 Inhomogeneous inequalities. Duality theorem. Existence theorem. We now turn again to the dual problems introduced at the beginning of this chapter, and we shall derive two fundamental theorems about them, using our results about homogeneous systems. Consider the system

$$-a_{11}y_1 - \ldots - a_{1m}y_m + c_1 t \geqslant 0,$$
$$\vdots$$
$$-a_{n1}y_1 - \ldots - a_{nm}y_m + c_n t \geqslant 0,$$

$$a_{11}x_1 + \ldots + a_{n1}x_n \qquad\qquad - b_1 t \geqslant 0,$$
$$\vdots \qquad\qquad\qquad\qquad\qquad \vdots$$
$$a_{1m}x_1 + \ldots + a_{nm}x_n \qquad\qquad - b_m t \geqslant 0,$$
$$-c_1 x_1 - \ldots - c_n x_n \qquad +b_1 y_1 + \ldots + b_m y_m \qquad \geqslant 0,$$
$$x_i,\ y_j,\ t \geqslant 0.$$

The matrix of coefficients of this system is skew-symmetric. Hence, according to Theorem 3–4, there exists a solution x_i^0, y_j^0, t^0 such that

$$-a_{i1}y_1^0 - \ldots - a_{im}y_m^0 + c_i t^0 + x_i^0 > 0 \qquad (i = 1, \ldots, n),$$
$$a_{1j}x_1^0 + \ldots + a_{nj}x_n^0 - b_j t^0 + y_j^0 > 0 \qquad (j = 1, \ldots, m),$$
$$-c_1 x_1^0 - \ldots - c_n x_n^0 + b_1 y_1^0 + \ldots + b_m y_m^0 + t^0 > 0.$$

In this solution the value of t^0 is either positive, or it is zero. If the value of t^0 is positive, then we divide the constraints by this value to find solutions to the dual problems. They are, moreover, optimal because we know already that for any solutions

$$\sum_j b_j y_j \leqslant \sum_j c_i x_i.$$

The particular solution just found satisfies also the reversed inequality, so that for this solution

$$\sum_j b_j y_j^0 = \sum_i c_i x_i^0.$$

Thus the x_i^0 as well as the y_j^0 produce the optimal value of their respective objective functions in the two problems. If t^0 is zero, then we have for solutions x_i, y_j of the dual systems

$$\sum_j b_j y_j^0 > \sum_i c_i x_i^0 \geqslant \sum_i x_i^0 \sum_j a_{ij} y_j = \sum_j (\sum_i a_{ij} x_i^0) y_j \geqslant 0,$$

and also

$$\sum_j b_j y_j^0 \leqslant \sum_j (\sum_i a_{ij} x_i) y_j^0 = \sum_i (\sum_j a_{ij} y_j^0) x_i \leqslant 0.$$

These two inequalities are contradictory, and hence at least one of the two dual systems has no solution in this case. It may be that neither has; for instance

$$x_1 - x_2 \geqslant 2, \qquad\qquad y_3 - y_4 \leqslant 1,$$
$$-x_1 + x_2 \geqslant -1. \qquad\qquad -y_3 + y_4 \leqslant -2.$$

Minimize $x_1 - 2x_2$.   Maximize $2y_3 - y_4$.

However, if one of the systems has a feasible solution, while the other has not, then the set of all feasible solutions of the former is unbounded, as will be shown.

Let $y_j = y_j^0$ $(j = 1, \ldots, m)$ be defined as above, and assume that the minimizing problem has no solution. Then $t^0 = 0$ and, as we have just seen, $\sum b_j y_j^0 > 0$. But when $t^0 = 0$, and if y_j' $(j = 1, \ldots, m)$ is a solution of the constraints of the maximizing problem, then so is $y_j' + r y_j^0$ whenever r is positive, and hence

$$\sum_j b_j(y_j' + r y_j^0) = \sum_j b_j y_j' + r \sum_j b_j y_j^0$$

can be made arbitrarily large.

A similar argument shows that $\sum_i c_i x_i$ can be made arbitrarily small, if there is no y_j to satisfy the constraints of the maximizing problem, while there is a solution to the minimizing one.

The results of these developments can be expressed in the following two statements.

DUALITY THEOREM. A feasible solution of one of two dual systems is optimal if and only if there exists a feasible solution to the dual problem such that the values of the objective function are equal.

EXISTENCE THEOREM. A problem—and hence its dual as well—has a (finite) optimal solution if and only if both have feasible solutions. If only one problem has a feasible solution, then its objective function is unbounded.

These theorems were first proved by D. Gale, H. W. Kuhn, and A. W. Tucker, in [72]. Another proof will be given in Chapter 5 in connection with the Simplex method. We state the consequences of these fundamental theorems in a form which is of importance from a computational point of view as well.

Let two dual systems be given, as follows:

$$\sum_i a_{ij} x_i \geqslant b_j, \qquad\qquad \sum_j a_{ij} y_j \leqslant c_i,$$
$$x_i \geqslant 0. \qquad\qquad\qquad y_j \geqslant 0.$$

Minimize $\sum_i c_i x_i$ Maximize $\sum_j b_j y_j$.

Let them have (finite) optimal solutions x_i^0, y_j^0. Then

$$\sum_i c_i x_i^0 = \sum_j b_j y_j^0,$$

while if x_i' and y_j' are solutions but not optimal solutions in their respective problems,

$$\sum_i c_i x_i' > \sum_j b_j y_j'.$$

Consequently the following system of inequalities is solved by the same values of the variables as are the two dual problems:

$$\sum_i a_{ij} x_i \geqslant b_j, \qquad\qquad \sum_j a_{ij} y_j \leqslant c_i,$$

$$x_i \geqslant 0, \qquad\qquad\qquad y_j \geqslant 0,$$

$$\sum_i c_i x_i \leqslant \sum_j b_j y_j,$$

where no objective function is explicitly mentioned.

Moreover, we have

$$\sum_i (c_i - \sum_j a_{ij} y_j^0) x_i^0 + \sum_j (-b_j + \sum_i a_{ij} x_i^0) y_j^0 = 0.$$

Since both factors in all the terms of the two sums on the left-hand side are, by assumption, non-negative, it follows that

$$x_i = 0 \qquad \text{for all } i \text{ for which } \sum_j a_{ij} y_j < c_i,$$

$$y_j = 0 \qquad \text{for all } j \text{ for which } \sum_i a_{ij} x_i > b_j,$$

and also

$$\sum_j a_{ij} y_j = c_i \qquad \text{for all } i \text{ for which } x_i > 0,$$

$$\sum_i a_{ij} x_i = b_j \qquad \text{for all } j \text{ for which } y_j > 0.$$

If one of the systems consisted entirely of equations, so that, for instance,

$$\sum_i a_{ij} x_i^0 = b_j,$$

then

$$\sum_i (c_i - \sum_j a_{ij} y_j^0) x_i^0 = 0,$$

and hence for each i we have either $x_i^0 = 0$, or $c_i = \sum_j a_{ij} y_j^0$. It is not necessary for y_j^0 to be non-negative to lead to this conclusion.

Consider now the *Lagrangian function* (the reason for this name will become clear in Chapter 12), defined as

$$\phi(x, y) = \sum_i c_i x_i + \sum_j b_j y_j - \sum_i \sum_j a_{ij} x_i y_j,$$

and again let x_i^0 and y_j^0 be optimal solutions of the two dual systems. Then

$$\phi(x^0, y) - \phi(x^0, y^0) = \sum_j (b_j - \sum_i a_{ij} x_i^0) y_j - \sum_j (b_j - \sum_i a_{ij} x_i^0) y_j^0 \leqslant 0,$$

because the subtrahend is zero. Similarly

$$\phi(x, y^0) - \phi(x^0, y^0) = \sum_i (c_i - \sum_j a_{ij} y_j^0) x_i - \sum_i (c_i - \sum_j a_{ij} y_i^0) x_i^0 \geqslant 0.$$

Concisely, we can write

$$\phi(x^0, y) \leqslant \phi(x^0, y^0) \leqslant \phi(x, y^0),$$

and we express this by saying that x^0, y^0 is a saddle point of the surface [in $(m + n)$-space] $\phi(x, y)$. (Note that if the inequality signs were reversed, we could still speak, symbolically, of a *saddle point*, but with different "orientation".)

This is a convenient place to establish a connection between linear programming methods and those of classical differential calculus (cf. [93] and [44]). Consider the following problem: Minimize $c_1 x_1 + \ldots + c_n x_n$ subject to

$$a_{1j} x_1 + \ldots + a_{nj} x_n \geqslant b_j \qquad (j = 1, \ldots, m),$$
$$x_i \geqslant 0 \qquad\qquad (i = 1, \ldots, n).$$

The constraints could be written

$$a_{1j} x_1 + \ldots + a_{nj} x_n - u_{n+j}^2 = b_j,$$
$$x_i - u_i^2 = 0.$$

The variables are all assumed to take only real values, so that the square of a variable is always non-negative. Now all constraints, without exception, are equations, and we can therefore apply the theory of Lagrange's multipliers. The latter will be denoted by $y_1, \ldots, y_{n+m}$, and it is then required to minimize

$$(c_1 x_1 + \ldots + c_n x_n) - \sum_{j=1}^{m} y_{n+j}(a_{1j} x_1 + \ldots + a_{nj} x_n - u_{n+j}^2 - b_j)$$
$$- \sum_{i=1}^{n} y_i (x_i - u_i^2).$$

Differentiating with respect to the variables x_i and u_i, we have

$$c_t - \sum_{j=1}^{m} y_{n+j} a_{tj} - y_t = 0 \qquad (t = 1, \ldots, n),$$

and

$$u_t y_t = 0, \qquad (t = 1, \ldots, n + m).$$

In other words, either $u_t = x_t = 0$, or

$$c_t - \sum_{j=1}^{m} a_{tj} y_{n+j} = 0, \qquad (t = 1, \ldots, n)$$

which is precisely the answer obtained earlier (apart from a slight change of notation). Similarly either $y_{n+s} = 0$ or

$$b_s - \sum_{i=1}^{n} a_{is} x_i = 0 \qquad (s = 1, \ldots, m).$$

It is of interest to note that if the Existence theorem is accepted as true, then Farkas' theorem can be derived from it, as follows: Consider the dual systems

$$a_{1j} x_1 + \ldots + a_{nj} x_n \geqslant 0 \qquad (j = 1, \ldots, m),$$
$$\text{Minimize } c_1 x_1 + \ldots + c_n x_n;$$

and

$$a_{i1} y_1 + \ldots + a_{im} y_m = c_i \qquad (i = 1, \ldots, n),$$
$$y_j \geqslant 0 \qquad\qquad\qquad (j = 1, \ldots, m),$$
$$\text{Maximize } 0 \cdot y_1 + \ldots + 0 \cdot y_m.$$

If for all solutions of the minimizing problem we have also $c_1 x_1 + \ldots + c_n x_n \geqslant 0$, then the minimum is obtained for $x_1 = \ldots = x_n = 0$, which is certainly feasible, whatever the values of the a_{ij}. But then, by the Existence theorem, the maximizing problem has also a solution: this is Farkas' theorem.

3-5 Orthogonality. The following aspect of duality was pointed out by A. W. Tucker (see pp. 33-34 in [17]). Consider the variables satisfying the constraints

(i) $a_{1j} x_1 + \ldots + a_{nj} x_n - x_{n+j} = b_j \qquad (j = 1, \ldots, m),$

and the variables satisfying the constraints

(ii) $y_i + a_{i1} y_{n+1} + \ldots + a_{im} y_N = c_i \qquad (i = 1, \ldots, n),$

where $n + m = N$.

Imagine that the $(x_1, \ldots, x_N)$ and $(y_1, \ldots, y_N)$ are coordinates of points in the same N-dimensional space. Then all points satisfying the set (i) form a space of $N - m = n$ dimensions, and all those satisfying (ii) form a space of m dimensions. These two spaces are orthogonal.

This is so, because any hyperplane of the set (i) is orthogonal to any hyperplane of the set (ii), since the inner product of the vector of coefficients of two such hyperplanes is zero. The coefficients are, respectively

$$a_{1j}, \ldots, a_{ij}, \ldots, a_{nj}, \qquad 0, \ldots, -1, \qquad 0, \ldots, 0,$$

and

$$0, \ldots, 1, \ldots, 0, \qquad a_{i1}, \ldots, a_{ij}, \ldots, a_{iN},$$

and their inner product is seen to be zero. Hence the respective intersections of the hyperplanes of the first set, and of the second set, are also orthogonal.

EXERCISES

3–1. Find a solution of

$$\begin{aligned} x_1 - 3x_2 + x_3 &= 0 & y_1 + y_2 &\geqslant 0 \\ x_1 - x_2 - x_3 &= 0 & -3y_1 - y_2 &\geqslant 0 \\ x_1, x_2, x_3 &\geqslant 0 & y_1 - y_2 &\geqslant 0 \end{aligned}$$

which satisfies also

$$\begin{aligned} x_1 + y_1 + y_2 &> 0 \\ x_2 - 3y_1 - y_2 &> 0 \\ x_3 + y_1 - y_2 &> 0. \end{aligned}$$

3–2. Find a solution of

$$\begin{aligned} 3x_1 + x_2 &= 0 & 3y_1 - 2y_2 &\geqslant 0 \\ -2x_1 + 2x_2 &= 0 & y_1 + 2y_2 &\geqslant 0 \\ x_1, x_2 &\geqslant 0 \end{aligned}$$

which satisfies also

$$\begin{aligned} x_1 + 3y_1 - 2y_2 &> 0 \\ x_2 + y_1 + 2y_2 &> 0. \end{aligned}$$

3–3. Show that all solutions of

$$\begin{aligned} 3x_1 + 2x_2 - x_3 &\geqslant 0 \\ 5x_1 - 3x_2 + x_3 &\geqslant 0 \\ 4x_1 - x_2 + 5x_3 &\geqslant 0 \end{aligned}$$

satisfy also $22x_1 + x_2 + 8x_3 \geqslant 0$, and hence find the non-negative coefficients mentioned in Farkas' theorem.

3–4. Prove that not all solutions of the first three inequalities in Exercise 3–3 satisfy $12x_1 + 7x_2 + 6x_3 \geqslant 0$.

CHAPTER 4

THEORY OF GRAPHS AND COMBINATORIAL THEORY

4–1 Definitions. Imagine points in a plane, called *nodes*, and lines connecting pairs of them, called *arcs*, or edges. Such a collection of nodes and arcs is called a *graph*. We consider two graphs to be equivalent if there is a one-to-one correspondence between their nodes and their arcs, in such a way that any arc joining two nodes corresponds to an arc joining the corresponding nodes.

A graph is called *planar* if no two arcs intersect, and if they meet only in common endpoints. Such a graph is, in particular, *a-b-planar* if it has an arc from node a to node b or, if it does not already contain such an arc, it can be drawn without intersecting any of the existing

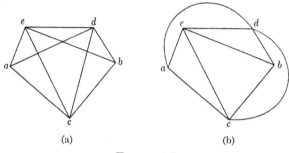

(a)   (b)

FIGURE 4–1

arcs. An a-b-planar graph is always equivalent to one which has the nodes a and b drawn, respectively, at the utmost left and the utmost right. To give an illustration, the graph of Fig. 4–1(a) is equivalent to that of part (b) which is a-d-planar, but not equivalent to an a-b-planar graph. More about these concepts will be found, for instance, in [4].

4–2 Shortest path. Let a graph be given and attach to each arc a number, its "length" from one of its nodes to the other. The lengths may be different for the two directions through an arc. It is required to find a *chain* (i.e. a succession of arcs without a gap between them) of minimal total length between two selected nodes, the *source* ⊕ and the *sink* ⊖. This problem can be formulated as one of linear programming, and this will be discussed in Chapter 6 (see Exercise 6–4). In the present chapter we exhibit a combinatorial method.

Attach to the source the number 0, and to all other nodes a very high number. As long as there is an arc from a to b, say, such that the number attached to b is larger than the total of the number attached to a and the length from a to b, reduce the number of b to be equal to this total. When no further reduction is possible, then trace a chain back from $\ominus$ to a point with a number equal to the number of $\ominus$ less the length of the arc from that point to $\ominus$, and so on further back until $\oplus$ is reached. The arcs through which we pass form the shortest chain. The proof of this is quite simple and is left to the reader.

EXAMPLE 4–1. See Fig. 4–2.* After a few simple steps we might

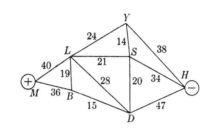

FIGURE 4–2

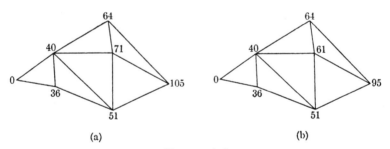

(a)   (b)

FIGURE 4–3

have reached, for instance, the diagram of Fig. 4–3(a), and after further reductions that of part (b). The shortest chain has a length of 95, and runs from M through L and S to H.

4–3 Maximal flow. Minimal cut. We consider again a graph with a source and a sink, and with numbers attached to the arcs as before, and

---

* Those familiar with the geography of England might be interested to know that the figures are statutory miles, and the letters are the initials of Barnsley, Doncaster, Hull, Leeds, Manchester, Selby, and York.

interpret them as *flow capacities* through the arcs. They may be different in the two directions. A graph with capacities is called a *network*.

A number attached to a chain from the source to the sink which is not larger than the capacity of any arc forming part of the chain is called a *chain flow*. A *network flow* is a collection of chain flows, such that the sum of the chain flows through any arc does not exceed its capacity.

If the sum of all chain flows through an arc equals its capacity, then we say the arc is *saturated*. The difference between its capacity and the chain flows through it is called its *residual flow*.

A collection of arcs such that each chain from the source to the sink contains at least one arc of the collection is called a *cut*; clearly no network flow can be larger than the sum of all capacities of the arcs of a cut. It follows that if a network flow can be found which equals the total of the capacities of the arcs of a cut, then that network flow is maximal, and the cut capacity is minimal. It will be shown that equality can always be achieved.

We can also impose limitations on the flow through nodes, but this does not introduce any essentially new aspect, because a node with a capacity c_i, say, can equivalently be represented as shown in Fig. 4–4.

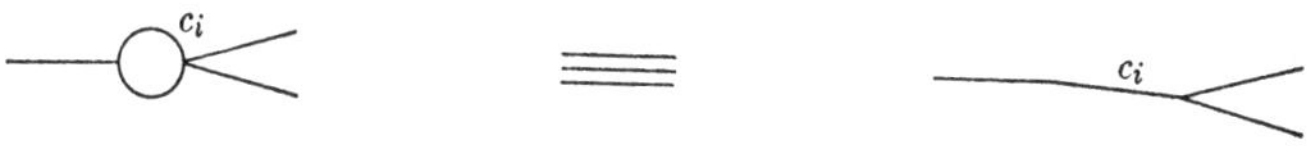

FIGURE 4–4

Let it be required to send the maximal flow from the source to the sink. This can be formulated as a linear programming problem as follows: Denote the flow from node i to node j by x_{ij}, and the capacity of that arc by r_{ij}. Let x_{ii} be the flow through node i, and r_{ii} its capacity. Any of these capacities may, of course, be infinite.

We identify the source by the first subscript 0, and the sink by the second subscript 0. Both the total flow out of the source, and into the sink, is x_{00}. Then the problem is to maximize x_{00} subject to

$$-x_{ii} + \sum_{j, j \neq i} x_{ij} = 0 \qquad (i = 0, 1, \ldots, n),$$

$$-x_{jj} + \sum_{i, i \neq j} x_{ij} = 0 \qquad (j = 0, 1, \ldots, n),$$

$$0 \leqslant x_{ij} \leqslant r_{ij} \qquad (i, j = 0, 1, \ldots, n).$$

This formulation is used in [50] to derive the equality of the maximal flow and the minimum cut capacity by a method using the Duality theorem. We shall here use a special method to construct the flow, and to derive the theorem.

We start off with some arbitrary network flow, for instance; a zero flow through all arcs (if we cannot think of anything better). Having found a network flow, there are two possibilities.

(i) There exists a chain from source to sink with no arc saturated. Then we increase the network flow by adding to each arc of such a chain the smallest residual flow of its arcs.

(ii) There is no chain from source to sink without a saturated arc in it. Then we have to find out whether a rearrangement of flows could serve to increase the network flow. This can be done by the following *labelling process*.

Label all nodes with an unsaturated arc from the source ⊕ by $(v, ⊕)$, where v is the residual flow in that arc. This indicates that a (further) flow v could be drawn from ⊕ into the node considered. Then

(a) if there remains any unlabelled node j with an unsaturated flow in an arc from a labelled node i to j, label the node j with (v, i), where v is now either the residual flow in that arc or the first number in the label of i, whichever is smaller. This is the (further) flow which could be drawn from i into j.

(b) if there remains an unlabelled node k from which a flow proceeds to a labelled node m, then label the node k by $(-u, m)$, where u is the first number in the label of m, or the flow from k into m, whichever is smaller. This indicates that a flow u could be withdrawn from m towards k.

It does not matter in which order one considers the two types of labelling mentioned in rules (a) and (b).

The aim is to find out whether the sink ⊖ can be labelled. Suppose this has been done, and the first number in its label is r. Then find a chain back from the sink to the source, guided by the nodes mentioned in the labels of the sink, of the node before it and so on. A flow can then be constructed by adding r to the flow through an arc in that chain, or subtracting r from the flow through such an arc, as indicated by the labels on its nodes. The total flow will thereby be increased by r.

If all labels have been applied by rules (a) and (b), and if B has remained without a label, then the maximal flow has been obtained, as will now be proved.

Let S be the set consisting of ⊕ and of all labelled nodes, and S' the set of all unlabelled nodes. Sink ⊖ belongs to S'. There is no flow away and out of S' from any of its nodes into one of S, since otherwise at least one of the nodes of S' could be labelled. Therefore the total flow equals the capacity of all arcs from a node of S to one of S'. These form a cut.

No flow can be larger than the capacity of a cut, and it follows that our procedure has produced a maximal flow, and also a minimal cut.

EXAMPLE 4–2. See Fig. 4–5. We can start with some obviously possible flow, but to proceed systematically, we start with the most

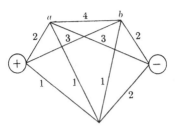

FIGURE 4–5

trivial one, viz. zero flow in all arcs. The labelling procedure then produces, in succession, graphs (a), (b), (c), (d), and (e) of Fig. 4–6. The total flows in these graphs are, respectively, 0, 2, 3, 4, and then 6. Thereafter the sink cannot be reached by labelling (actually all nodes

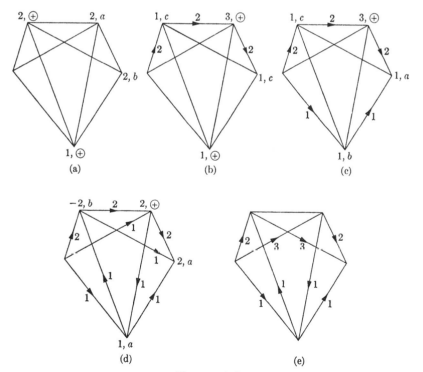

FIGURE 4–6

except the source belong to S'), and the maximal flow has been reached. In this case a cut is given by the arcs emanating from the source ⊕.

Extensive problems do not lend themselves readily to a graphical representation, and we resort then to an equivalent tabular method.

We use a table in which rows refer to nodes from which a flow can emanate, and columns to nodes into which it can flow. To begin with, we have in the various cells the capacities of the routes, possibly different in the two directions. When some flow has been established, say from a to b, then we subtract the amount of the flow from the entry in cell ab, so that the residual flow remains there. Also, in cell ba we add to the entry the amount of the flow. We may think of this as an additional capacity, in that the flow from a to b can be withdrawn, or redirected to flow from b to a, cancelling the previous flow. This has the effect that the flow from a to b is at any time given by the difference between the original entry in that cell and the present number, or also by the negative difference of the original entry in the cell ba (symmetrical to ab) and its present entry. Labelling proceeds as before, though it is unnecessary to use the minus sign.

This will be illustrated by the same example as was just used. Arcs which do not exist are listed as having capacity 0. The example is symmetrical in its capacities, but this is irrelevant.

EXAMPLE 4–3. The table obtained is as indicated:

	⊕	a	b	c	⊖	
⊕	–	2^-	3	1	0	
a	2^+	–	4^-	1	3	$(2, ⊕)$
b	3	4^+	–	1	2^-	$(2, a)$
c	1	1	1	–	2	$(1, ⊕)$
⊖	0	3	2^+	2	–	$(2, b)$

The flow at present is zero. The labels are the same as those in the graphical presentation. The sink has been labelled, and the resulting additional flow of $r = 2$ leads to the indicated additions and subtractions of 2. The resulting flow of 2 is given in the following table:

	⊕	a	b	c	⊖	
⊕	–	0	3	1^-	0	
a	4	–	2	1	3	$(1, c)$
b	3	6	–	1	0	$(3, ⊕)$
c	1^+	1	1	–	2^-	$(1, ⊕)$
⊖	0	3	4	2^+	–	$(1, c)$

There are various possibilities of labelling, but we choose to proceed in parallel with the graphical version of this example.

⊕	a	b	c	⊖	
⊕	–	0	3^-	0	0	
a	4	–	2	1^+	3^-	$(1, c)$
b	3^+	6	–	1^-	0	$(3, ⊕)$
c	2	1^-	1^+	–	1	$(1, b)$
⊖	0	3^+	4	3	–	$(1, a)$

⊕	a	b	c	⊖	
⊕	–	0	2^-	0	0	
a	4	–	2^+	2	2^-	$(2, b)$
b	4^+	6^-	–	0	0	$(2, ⊕)$
c	2	0	2	–	1	$(1, a)$
⊖	0	4^+	4	3	–	$(2, a)$

Finally

⊕	a	b	c	⊖
⊕	–	0	0	0	0
a	4	–	4	2	0
b	6	4	–	0	0
c	2	0	2	–	1
⊖	0	6	4	3	–

The flow, for instance, from ⊕ to b is 3, since this is the difference between the first and the last entry in cell ⊕b. The withdrawal of a flow from a to b (its reversal), indicated by the negative label in the graph, is now shown in the fact that the entry in cell ab, previously decreased from 4 to 2, has in the last table again increased to 4.

The labelling method, whether graphical or tabular, can be applied to a graph whether it is planar or not. However, for a ⊕-⊖-planar graph a simpler method is available. Imagine the graph (or one equivalent to it) drawn in such a way that the source is on the extreme left, and the sink on the extreme right. Consider then the uppermost chain from the source to the sink or, to express it differently, turn at every junction to your left, returning if you reach a dead end. Send the maximal flow through this chain, note it, and subtract then the flow from all arcs of the chain. Delete arcs which were saturated in the chain; there will be at least one. Repeat the procedure until no further chain exists from ⊕ to ⊖. The maximal network flow is then the sum of all the chain flows thus constructed. The same argument as before shows that we have indeed reached the maximal flow, and have found a minimal cut of the same total capacity.

EXAMPLE 4–4. See Fig. 4–7. The total flow out of A, that into B, and that through the arcs of a minimal cut (marked III) is 95.

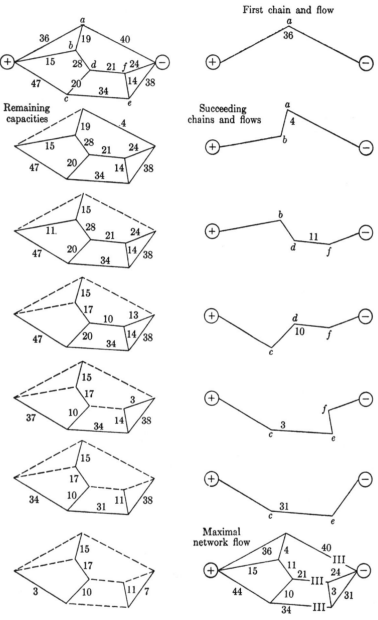

FIGURE 4-7

The developments of this section are based on various reports of the RAND Corporation, and in particular on [65].

A powerful method for finding the shortest path from a given origin to all other nodes in the network has been communicated by G. B. Dantzig in [47].

4–4 Dual graphs. When an a-b-planar network with source a and sink b is given, then its dual network is defined by the following construction. Let the source and the sink be, respectively, the leftmost and the rightmost points in the graph. Draw arcs with capacity zero from the source to the left to infinity, and from the sink to the right to infinity. The network remains planar, and its graph partitions the plane. Mark one point in each area and connect it with the points of adjacent areas by not intersecting arcs. Assign to each of these arcs the number of that arc of the first network which it crosses. Call the points in the unbounded portions source and sink, respectively. The resulting network is the dual of the original one, and the relationship of duality is symmetrical.

Let two dual networks N_1 and N_2 be given. If we take a cut in N_1, then the corresponding arcs of N_2 form a connected network, i.e. a chain from its source to its sink exists. A minimal cut of N_1 corresponds to a chain of smallest length from source to sink in N_2. Hence that smallest length equals the maximal flow of the dual networks. (The graphs in Examples 4–1 and 4–4 are dual.) The shortest chain in the first has a length of 95, and the maximal flow through the second is 95, which is also the total capacity of its minimal cut.

4–5 Directed network of the transportation problem. An arc of a network with a positive capacity in one direction only is called *directed*, and a *directed network* is one in which all arcs are directed. Such a network of special interest is that which represents a transportation problem (see Chapter 1). It consists of two sets, S and T, of nodes, and there are only arcs connecting a node of S with one of T, and with capacities only in this direction. These capacities may all be infinite. If they are not, then we speak of a *capacitated* network.

We relate the nodes of S to sources, and those of T to destinations in a transportation problem. In order to record the availabilities and the requirements, we add to the graph a master source $\oplus$, with arcs to all nodes of S, having capacities away from $\oplus$ equal to the several availabilities, and also a master sink $\ominus$, with arcs directed from the nodes of T to it, and having capacities equal to the several requirements. The arcs from nodes of S to nodes of T have unlimited capacities in the basic form of the transportation problem.

In such a problem we have also costs connected with the various routes. We deal with these in Chapter 6. Here we consider the problem of sending the maximal flow from $\oplus$ to $\ominus$. This is a trivial problem, unless the arcs from S to T are capacitated.

We can, of course, solve such a problem by the general method. But if we used the tabular presentation, then many of the capacities would be listed as zero (e.g. those from a node of T to one of S), and it is possible to collapse the table and simplify the labelling. We illustrate this by an example. The reader will not have any difficulty in recognizing the parallelism with the general procedure, and in convincing himself thereby of the validity of the simplification.

EXAMPLE 4–5. We have

		d	e	f	
		3	5	5	
		1)	1)	2)	
a	2	1	1		(1, d)
		2)	1)	2)	
b	2	2			(1, d)
		3)	5)	4)	
c	9		4	4	(1, $\oplus$)
		(1, c)	(1, c)	(1, a)	

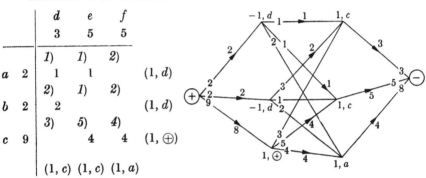

The numbers such as *1)* are the capacities of the routes. The sources are a, b, and c, and the destinations are d, e, and f. We have entered a possible flow and also the first labels. There is a deficiency of 1 in row c. In this row, there are residuals in columns d and e. Then, in column d, we look for flows and label the rows in which they appear. In row a there is a residual in column f, still unlabelled, and therefore a label $(1, a)$ can be attached. Finally, since we have now labelled a column with a deficiency, an increase can be made in the total flow, and we obtain

		d	e	f
		3	5	5
		1)	1)	2)
a	2		1	1
		2)	1)	2)
b	2	2		
		3)	5)	4)
c	9	1	4	4

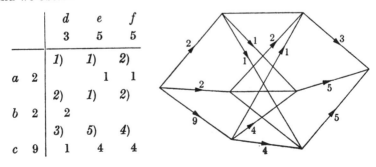

No deficiency is left in any row or column, and we have reached the maximal flow.

For easy reference, we mention here the rules for the case where the only capacity limits are zero (i.e. no arc) or infinity, quoting—freely— from [66].

The labelling process is a way of alternately assigning labels first to certain rows, then to certain columns, then further rows, etc. Start by labelling those rows that have surpluses. Attach the label $(s, \oplus)$ to each such row, where s is the surplus in that row. Here s will be referred to as the potential flow of the row.

Take each row in turn and sweep along it looking for possible flows in columns not previously labelled. Label the column of such a possible flow (m, R), where m is the potential flow of the row being scanned, and R identifies that row. The amount m will be called the potential flow of the column. The process is continued until all the rows have been swept, unless a column is labelled which has a shortage, in which case a *breakthrough* has been achieved.

Having scanned each labelled row, we then scan the newly labelled columns looking now for positive flows not in any of the previously labelled rows; the row in which such a flow lies is given a label (n, C), where C identifies the column that is being scanned and n is the smaller of the potential flow of the column and the flow in the route. The quantity n will be called the potential flow of the newly labelled row.

Having scanned all the newly labelled columns, we turn again to the rows, scanning the newly labelled ones as above. The second entry in the label, which was $\oplus$ above, will now refer to the column of surpluses.

In the case of a breakthrough we have labelled a column with a shortage. Suppose the label is (m, R), and the shortage is s. Then increase the flow in the cell where row R meets this column by the smaller of m and s. Decrease the entry in the cell where the column designated in the label of row R meets that row by the same amount, obtain a new row from the label of the column just dealt with and increase, and so on, until a row is reached whose label contains $\oplus$. Then start labelling again.

If no column with a deficiency can be labelled, either because there is no such column left, or because such a column cannot be reached by a label, then the maximal flow has been obtained.

4-6 Trees. Triangularity.

The determination of maximal flow through a network will be of use when we deal with the Transportation Problem in Chapter 6. Here we introduce now further graph-theoretical considerations which have also connections with that problem.

A graph is called a *tree* if each node can be reached from any other by a chain between them, and if there is no *loop* in it, i.e. no chain from any point back to itself.

If there are p nodes in a tree, single out one node and call it the *root*. Taking any other of the remaining nodes, consider the chain from the root to it. There cannot be more than one chain, since otherwise arcs of two of them would form a loop. Hence there will be precisely one arc terminating in each of the $p - 1$ nodes, and there will be precisely $p - 1$ arcs in the tree. If there are p or more arcs in a graph of $p - 1$ nodes, apart from the root, then there must be a loop somewhere in it.

We establish now a connection between trees and a system of linear equations of a type called *triangular*. A triangular set of p equations in p unknowns $x_1, \ldots, x_p$ is one which contains an equation in just one single variable, then an equation containing perhaps the latter, but containing one more, and so on, so that there is always an equation which contains merely one unknown which did not appear in an earlier equation. For instance,

$$
\begin{aligned}
x_1 &= 1 \\
x_1 + 3x_2 &= 2 \\
x_1 + 2x_2 + 3x_3 &= 5 \\
4x_2 \qquad + x_4 &= 2 \\
2x_1 \qquad + 2x_3 \qquad + x_5 &= 4
\end{aligned}
$$

is a triangular set, and the way it is here written explains the name. Such a set can be easily solved by solving one equation at a time and substituting the obtained values in the next. Thus in the present example, we have $x_1 = 1$. Then $1 + 3x_2 = 2$, i.e. $x_2 = 1/3$. Continuing, $x_3 = 10/9$, $x_4 = 2/3$, $x_5 = -2/9$.

Consider now a tree of p nodes, and hence of $p - 1$ arcs, and attach a value to each arc. We can then give an arbitrary value to some node, chosen to be the root, and proceed from the root to any node in a succession of arcs, attaching to each node encountered a value such that the two nodes at the end of each arc have values which add up to the value of the arc. The equations to determine the values of the nodes will be triangular, and there will be no contradiction, because each node can only be reached in a unique way.

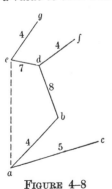

FIGURE 4-8

EXAMPLE 4-6. In the graph of Fig. 4-8 we make arbitrarily $x_a = 0$ and obtain then

i.e.

$$x_b = 4, \qquad x_b + x_d = 8, \qquad x_d + x_f = 4,$$
$$x_c = 5, \qquad x_d + x_e = 7, \qquad x_e + x_g = 4,$$

$$
\begin{array}{ccccccc}
x_a & x_b & x_c & x_d & x_e & x_f & x_g \\
0 & 4 & 5 & 4 & 3 & 0 & 1.
\end{array}
$$

Now add the arc ae, thereby creating a loop. We then also have a value for $x_a + x_e$. It will either contradict the previous results, or confirm it, without adding any new information. Either fact indicates the occurrence of a loop in the graph. We can find it by tracing our steps back through the two paths which ended at a point whose value was already known. By dropping, in our example, any one of ab, bd, de, or ae, the loop disappears, and we have once more a tree.

4–7 Incidence matrix. It is often convenient to describe a graph in tabular form. To this end consider a matrix (d_{ij}), whose columns correspond to the arcs of a graph, and whose rows correspond to the nodes. If the arc of the jth column goes from the node i_1 to the node i_2, then we write $d_{i_1 j} = 1$, and $d_{i_2 j} = 1$. All other entries in the jth column are zero. We assume that there is no row consisting entirely of zeros; otherwise we would have a node unconnected with all others.

The matrix just described is called an *incidence matrix*. Let its rank be r. We choose then r columns which are linearly independent, i.e. such that all the other columns can be expressed as a linear combination of them. We call these r columns a *basis* (of the incidence matrix).

A basis must include at least one column with a 1 in any given row, since otherwise a column with such an entry in that row could not be expressed by columns in the basis.

Consider now a tree in the graph which contains all nodes. Using the arcs of the tree, we can go from any node to any other. Selecting an arc that does not belong to the tree, we can proceed from its starting node to its end node by a chain in the tree, and if we take the column corresponding to the first arc, subtract that of the second, and add and subtract alternately in this way, then we obtain the column of the arc that we have selected.

EXAMPLE 4–7. In Fig. 4–9 the graph G contains the tree T. The incidence matrix of G is

$$
\begin{array}{cccccccc}
a & 1 & 1 & 1 & 0 & 0 & 0 \\
b & 0 & 0 & 0 & 1 & 1 & 1 \\
c & 1 & 0 & 0 & 1 & 0 & 0 \\
d & 0 & 1 & 0 & 0 & 1 & 0 \\
e & 0 & 0 & 1 & 0 & 0 & 1
\end{array}
$$

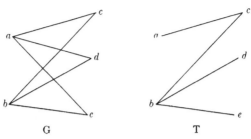

FIGURE 4–9

A basis, corresponding to T, is

$$\begin{matrix} 1 & 0 & 0 & 0 \\ 0 & 1 & 1 & 1 \\ 1 & 1 & 0 & 0 \\ 0 & 0 & 1 & 0 \\ 0 & 0 & 0 & 1 \end{matrix}$$

The other columns can be represented as follows:

$$ad = ac - bc + bd \qquad \text{and} \qquad ae = ac - bc + be.$$

Let the matrix

$$\begin{pmatrix} a_{11} & \cdots & a_{n1} \\ \cdot & & \\ \cdot & & \\ \cdot & & \\ a_{1m} & \cdots & a_{nm} \end{pmatrix} \qquad (m \leqslant n)$$

have rank r. There exists then a submatrix, say

$$\begin{pmatrix} a_{11} & \cdots & a_{r1} \\ \cdot & & \\ \cdot & & \\ \cdot & & \\ a_{1m} & \cdots & a_{rm} \end{pmatrix},$$

such that any other column of the first matrix can be represented as a linear combination of the first r, i.e. values $t_1, \ldots, t_r$ can be found such that, taking the $(r + 1)$th column as an example,

$$a_{r+1,j} = t_1 a_{1j} + \ldots + t_r a_{rj} \qquad (j = 1, \ldots, m).$$

Now consider the system of linear equations

$$a_{1j}x_1 + \ldots + a_{nj}x_n = b_j \qquad (j = 1, \ldots, m).$$

If we solve it for $x_1, \ldots, x_r$, then the coefficient of x_{r+1} in the expression of x_j, say, is the negative of t_j which appears as a solution to the previous problem.

EXAMPLE 4–8. Take the incidence matrix of Example 4–7 to be the matrix of coefficients of a system of linear equations:

$$
\begin{aligned}
x_{11} + x_{12} + x_{13} & & & = a_1, \\
& x_{21} + x_{22} + x_{23} & & = a_2, \\
x_{11} & + x_{21} & & = b_1, \\
x_{12} & + x_{22} & & = b_2, \\
x_{13} & + x_{23} & = b_3.
\end{aligned}
$$

This is, of course, a transportation problem. If we solve this system for x_{11}, x_{21}, x_{22}, and x_{23}, we have

$$
\begin{aligned}
x_{11} &= a_1 & - x_{12} - x_{13}, \\
x_{21} &= b_1 - a_1 + x_{12} + x_{13}, \\
x_{22} &= b_2 & - x_{12}, \\
x_{23} &= b_3 & - x_{13}.
\end{aligned}
$$

The coefficients of x_{12} in these four equations are the negatives of those in the expression of ad. Those of x_{13} are the negatives of the coefficients in the expression of ae.

4–8 Dantzig property. Unimodular property. In this section we give a more systematic account of properties of incidence and other matrices and, in particular, of the matrix of the Transportation Problem. The development is mainly based on two papers, by A. J. Hoffman and J. B. Kruskal [87], and by I. Heller and C. B. Tompkins [82].

We have just seen that if we solve a system of equations whose matrix of coefficients is an incidence matrix of rank r for r variables, then the coefficients of the other variables in the solution will be 0, 1, or -1. No other coefficient will appear. Heller and Tompkins call this the *Dantzig property*, because it was first demonstrated, in connection with the Transportation Problem, by G. B. Dantzig in [40].

In a transportation problem solutions will, very frequently, only have a realistic meaning if the values of the variables are integers. It is therefore of interest to investigate conditions which ensure this to be the case.

Heller and Tompkins have proved the following theorem.

THEOREM 4–1. Let the matrix of coefficients of a linear programming problem $\sum_i a_{ij}x_i = b_j$ have the Dantzig property, and let there exist a solution in which all x_i have integral values, say $x_i = x_i^0$. Then all basic solutions have integral coordinates.

Proof. We have

$$\sum_{i=1}^{N} a_{ij}x_i = b_j \qquad (j = 1, \ldots, m).$$

If, then, $(x_1', \ldots, x_m', \ 0, \ldots, 0)$ is a basic solution, then

$$\sum_{i=1}^{m} a_{ij}x_i' = b_j \qquad (j = 1, \ldots, m).$$

Because of the Dantzig property, we have

$$a_{ij} = c_{i1}a_{1j} + \ldots + c_{im}a_{mj} \qquad (i = m + 1, \ldots, N),$$

where the $c_{i1}, \ldots, c_{im}$ are all integers (in fact, they are 1, -1, or 0). By substitution we obtain

$$a_{1j}(x_1^0 + c_{m+11}x_{m+1}^0 + \ldots + a_{N1}x_N^0) + \ldots$$
$$+ a_{mj}(x_m^0 + c_{m+1m}x_{m+1}^0 + \ldots + c_{Nm}x_N^0) = b_j \qquad (j = 1, \ldots, m).$$

Now

$$\sum_{i=1}^{m} a_{ij}x_i' = b_j$$

has a unique solution, so that

$$x_i' = x_i^0 + c_{m+1i}x_{m+1}^0 + \ldots + c_{Ni}x_N^0.$$

The right-hand sides of all these expressions have integer values, which proves the theorem.

Of greater significance is the *unimodular property* (so-called in [87]) that not merely all single elements, but all minors of the matrix have values 0, 1, or -1.

If all right-hand sides have integer values, then it follows from the solution of the system by determinants (Cramer's rule, see the Appendix) that all basic solutions have integral values. The converse, i.e. that if all basic solutions consist of integral values, then the matrix has the unimodular property, is proved in [87].

This paper contains various other sufficient conditions for the same property to hold, for instance that of the following theorem.

THEOREM 4–2. An incidence matrix of a directed graph (i.e. such that the arcs can only be traversed in a given direction) has the unimodular property if and only if all its loops can only be traversed by going alternately with and against the direction of its arcs.

It follows that the matrix of a transportation problem is unimodular. No necessary and sufficient condition is known for a general matrix (not necessarily one of incidence) of zeros and ones to have the unimodular property.

The following fundamental theorems are contained in the Appendix to [82].

THEOREM 4–3*. Let A be a matrix whose rows can be partitioned into two disjoint sets B and C, such that:

(1) Every column contains at most two nonzero entries.
(2) Every entry is 0, 1, or −1.
(3) If two nonzero entries in the same column have the same sign, then the row of one is in B and that of the other in C; if they have opposite signs, then both are in B or both are in C.

Then A has the unimodular property. The matrix of the Transportation Problem is of this type.

THEOREM 4–4†. If a matrix A satisfies (1) of Theorem 4–3, and has the unimodular property, then (2) and (3) of that theorem are also satisfied. From this we can deduce that every unimodular matrix which satisfies (1) is of the transportation type.

To begin with, if there exist columns with only one nonzero entry, then for each such column we add a row, with another nonzero entry in that column only, without destroying the unimodular property. (For instance, by repeating the entry with opposite sign.)

Then, by changing the signs in all rows belonging to C, say, each column will contain precisely one entry 1, and one entry −1.

To transform such a matrix into one of the transportation type, we consider the columns of coefficients to refer to routes, and the rows to ports of departure when the entry is 1, and to ports of arrival when it is −1. Some ports may appear in both capacities (*transshipment*, see Section 6–2).

---

* Proved by Heller and Tompkins and, by another method, by A. J. Hoffman.
† Proved by D. Gale.

The equations of the Transportation Problem are not independent, so that the right-hand sides are subject to obvious conditions for the system to be feasible.

EXAMPLE 4–9.

(a) $x_1 - x_2 \quad = \quad 6$

(b) $\quad\quad x_2 + x_3 = \quad 3$

(c) $-x_1 \quad\quad - x_3 = -9$

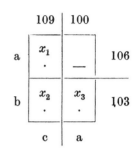

4–9 Systems of distinct representatives. Related theorems. We shall now deal with a combinatorial problem of great interest in itself, but we introduce it by relating it to a problem of planning. Imagine that n persons have applied for m jobs and that it has been decided for each of them for which jobs he is suitable, and for which he is not. It is required to determine whether it is possible to fill each job with a suitable person, each person filling only one single job. Clearly $n \geqslant m$ is a necessary condition. We shall derive sufficient conditions.

Construct a table $a_{ij} = 1$ if person i is suitable for job j, and $a_{ij} = 0$ otherwise. Let $x_{ij} = 1$ if person i fills job j, and $x_{ij} = 0$ otherwise. We have

$$\sum_{j=1}^{m} x_{ij} \leqslant 1 \quad\quad \text{for all } i,$$

(No person can have more than one job, but he might not get any.) and

$$\sum_{i=1}^{m} x_{ij} \leqslant 1 \quad\quad \text{for all } j.$$

(No job can be filled by more than one person, but it might not be filled at all.) We want to fill as many jobs as possible with suitable candidates, i.e. we want to maximize

$$\sum_i \sum_j a_{ij} x_{ij}.$$

If this maximum is m, then all jobs can be filled. We are, thus, looking for conditions which assure that the maximum equals m.

If we attach to each job j the subscripts i for which $a_{ij} = 1$, then we may call the set of these subscripts the set S_j, and any i in it a representative of the set. Looking for persons to fill all jobs can then be expressed as looking for a *set of distinct representatives* (SDR) of the m given sets.

A sufficient and necessary condition for such a SDR to exist is, as we have said, that the maximum of the linear programming problem above should be m. To transform this into simpler conditions, we consider the dual to that problem, viz. minimize $\sum_i u_i + \sum_j v_j$, subject to

$$u_i + v_j \geqslant a_{ij} \quad (i = 1, \ldots, n;\ j = 1, \ldots, m),$$
$$u_i \geqslant 0, \quad v_j \geqslant 0 \qquad \qquad \text{(all } i, j\text{)}.$$

It follows from the form of the constraints that in the optimal solution all u_i and v_j are either 0 or 1, and if $a_{ij} = 1$, then either u_i or v_j must be unity.

Hence the dual problem can be phrased as follows: Choose the smallest number of elements from $1, 2, \ldots, n$ and sets from $S_1, \ldots, S_m$ so that, if i is in S_j, then either i or S_j is chosen.

Assume now that $\sum_i u_i + \sum_j v_j$ is smaller than m. Then there will be sets which were not chosen, say k in number, and they cannot contain k distinct representatives between them, because if they did, then they would make up the number m. It follows that if any k $(1, 2, \ldots, m)$ sets S_j have k distinct representatives between them, then this is sufficient for the minimum of $\sum_i u_i + \sum_j v_j$, and hence for the maximum of $\sum_i \sum_j a_{ij} x_{ij}$, to be m. Clearly this condition is necessary, and we have thus proved *Hall's theorem* [80]: There exists a set of distinct representatives of m sets if and only if any k of the sets contain between them at least k distinct elements, where $1 \leqslant k \leqslant m$.

This proof is based on that in the Introduction to [11]. Of immediate interest to our subject is the fact that Hall's theorem is equivalent to a theorem by D. König, in the sense that either can be derived from the other. We show this now, following the exposition in [79].

We start from Hall's theorem which we have just proved. Let a $n \times t$ matrix of 0 and 1 be given, such that the smallest number of lines (rows or columns) containing all zeros is m, and the largest number of independent zeros (i.e. not two of them being in the same line) is M.

Because each line contains only one independent zero at the most, we have $m \geqslant M$. *König's theorem* states that $m = M$. To prove it, it is sufficient to show that $m \leqslant M$.

Let the m lines which contain all zeros be the first s columns (without restriction of generality) and r of the rows. Define the set S_j $(j = 1, 2, \ldots, s)$ as the set of the subscripts of those rows which contain zeros in column j, but excluding those rows which are amongst the r mentioned.

Illustration.

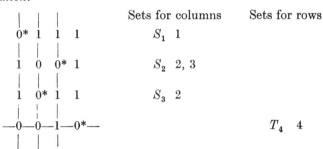

					Sets for columns		Sets for rows
0*	1	1	1		S_1	1	
1	0	0*	1		S_2	2, 3	
1	0*	1	1		S_3	2	
—0—	0—	1—	0*—				T_4 4

If k of these sets contained less than k zeros between them, then the corresponding rows could be used to replace the columns corresponding to the k sets. But this is impossible if m is the minimum number of covering lines. We can therefore find s independent zeros in those s columns, and none of these being in the r rows used for cover. They are marked by an asterisk.

Apply the same argument to rows. Thus there exist at least $r + s = m$ independent zeros and hence $m \leqslant M$. Therefore $m = M$. This is *König's theorem* [9, p. 240].

Given the latter, P. Hall's theorem can be derived from it as follows: The necessity of the condition is obvious, as has already been mentioned. If a minimal covering contains $r + s = m$ lines, then by König's theorem there are m independent zeros in the matrix, and the corresponding subscripts of the rows and columns form a SDR for the m sets (for this purpose we might number the rows from 1 to n, and the columns from $n + 1$ to $n + m$). This completes the proof of the equivalence of the two theorems. König's theorem will again be referred to in Chapter 6.

Hall's theorem has been introduced at this stage as answering a problem of feasibility. It answers also the following problem, again one of feasibility (cf. [71]): If one unit of an article is produced at each of a set of origins, and one unit is required at each of a set of destinations, when is it possible to supply each destination with its required units? The answer is that this is feasible if and only if each set of k destinations is connected with at least k origins. Gale has generalized this result to the case of general (not necessarily unit) supplies, and to capacity restrictions on the routes.

A further theorem, related to problems of Linear Programming, will now be derived from Hall's theorem. We shall prove that if we have a square $n \times n$ table of non-negative values such that the rows and the columns all add up to the same positive constant t, then this table can be written as a sum of *permutation matrices*, with positive coefficients. A permutation matrix is a square table with all entries 0 or 1, the latter in such positions that there is just one of them in each row and one in each column. It arises from the unit matrix

$$\begin{pmatrix} 1 & 0 & \ldots & 0 \\ 0 & 1 & \ldots & 0 \\ \cdot & & & \\ \cdot & & & \\ \cdot & & & \\ 0 & 0 & \ldots & 1 \end{pmatrix}$$

by a permutation of rows, or columns, or both.

Let the table be (x_{ij}) $(i, j = 1, \ldots, n)$ and let the number of non-zero elements be w. Clearly, $w \geqslant n$, and if $w = n$ then the table is t times a permutation matrix, i.e. one which arises from

$$\begin{pmatrix} t & 0 & \ldots & 0 \\ 0 & t & \ldots & 0 \\ \cdot & & & \\ \cdot & & & \\ \cdot & & & \\ 0 & 0 & \ldots & t \end{pmatrix}$$

by row and/or column permutations.

Let S_i be the set of those j's for which $x_{ij} \neq 0$. If the condition in Hall's theorem were not satisfied, then some set of k rows would have their nonzero elements in less than k columns, and then the sums of those rows would add up to kt, but the set of columns, less than k in number, could not, which contradicts the original assumptions. Hence distinct representatives of the sets S_i can be found. Let them be $j_1, \ldots, j_n$. Then (P_{ij}) is a permutation matrix, if $p_{ij_i} = 1$ and all other p_{ij} for a given i are zero.

Let u be the smallest of the x_{ij_i}. Then the matrix $(x_{ij} - up_{ij})$ has more zero elements than (x_{ij}), but the two matrices are of the same type, only that columns and rows now add up to $t - u$. Our statement can thus be proved by complete induction.

This applies, in particular, to *doubly stochastic matrices*, i.e. matrices where each row and each column has non-negative entries that add up to unity.

EXAMPLE 4–11.

$$\begin{pmatrix} 0.5 & 0.3 & 0.2 \\ 0.2 & 0.2 & 0.6 \\ 0.3 & 0.5 & 0.2 \end{pmatrix} = 0.2\begin{pmatrix} 0 & 0 & 1 \\ 0 & 1 & 0 \\ 1 & 0 & 0 \end{pmatrix} + 0.2\begin{pmatrix} 0 & 1 & 0 \\ 1 & 0 & 0 \\ 0 & 0 & 1 \end{pmatrix} +$$

$$+ 0.1\begin{pmatrix} 0 & 1 & 0 \\ 0 & 0 & 1 \\ 1 & 0 & 0 \end{pmatrix} + 0.5\begin{pmatrix} 1 & 0 & 0 \\ 0 & 0 & 1 \\ 0 & 1 & 0 \end{pmatrix}.$$

Hence, if x_{ij} $(i, j = 1, \ldots, n)$ are subject to the constraints $\sum_i x_{ij} = 1$, $\sum_j x_{ij} = 1$, then we can find values (weights) w_s such that

$$x_{ij} = \sum_s w_s p_{ij}^s \qquad \text{(all } i, j),$$

where all (p_{ij}^s) are permutation matrices.

On the other hand, a permutation matrix can clearly not be a combination of others. Hence vertices of the feasible region of a linear programming problem with the above constraints (cf. The Assignment Problem, Section 6–6) correspond to permutation matrices. In view of Theorem 2–3 it follows that there exists always an optimal solution to a problem with these constraints which is a permutation matrix; no fractional values of the x_{ij} need ever be considered.

The following consequence of the theorem about matrices with constant row and column sums may here be mentioned. Let there be in a crowd n girls and n boys. Each boy has m girl friends, and each girl has m boy friends. Is it possible to divide the crowd into couples in such a way that each girl goes with one of her boy friends, and each boy with one of his girl friends? The affirmative answer follows, if we construct a matrix, with rows for boys and columns for girls, and an entry 1 for each boy-friend/girl-friend combination, while all other entries are zero. (In slightly different contexts, this example appears in [9, p. 175] and also in [4].)

EXERCISES

4–1. Solve Example 4–4 by the tabular method of labelling.

4–2. Find a set of distinct representatives of the following sets:

(a) 1, 2, 3, 4, 5   (e) 3, 4
(b) 2, 3, 4   (f) 2, 4
(c) 1, 3, 4, 5   (g) 1, 4, 6, 9
(d) 1, 3, 4   (h) 2, 5, 7, 8, 9

If this is not possible, which extension of one of the sets would make it possible ?

4–3. In the graph of Fig. 4–10, the short lines have "length" 1, and the long ones "length" 2. Find the shortest route from A to B. Draw the dual graph and determine the maximum flow through it.

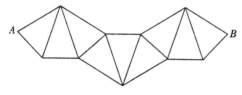

FIGURE 4–10

4–4. Find all sets of independent "1"s in the following table.

$$
\begin{array}{ccccc}
0 & 1 & 1 & 1 & 0 \\
1 & 0 & 1 & 0 & 1 \\
1 & 1 & 0 & 1 & 0 \\
0 & 1 & 0 & 1 & 1 \\
1 & 0 & 1 & 0 & 1
\end{array}
$$

(Note that all the row sums and all the column sums are 3.)

CHAPTER 5

GENERAL ALGORITHMS

We shall now consider in detail computational methods for the solution of linear programming problems.

5-1 Simplex method. To begin with, we deal with the Simplex method, due to George B. Dantzig, who described it first in [39]. This method, with some modifications, is still the one most frequently used for linear programming computations.

The Simplex method deals with constraints written as equations in non-negative variables, and it considers only basic feasible solutions. We have seen in Section 2–2 that if optimal feasible solutions exist at all, then a basic feasible optimal solution also exists. Hence no essential point is lost if we deal with basic feasible solutions only, provided there is a criterion which allows us to recognize it if other, basic or nonbasic optimal solutions exist as well.

To begin with, we introduce a convenient shorthand presentation. Let a basic solution be such that $x_1, \ldots, x_m$ are the basic variables with values $x_1^0, \ldots, x_m^0$, and $x_{m+1}^0 = \ldots = x_N^0 = 0$. Then the constraints can be transformed into

$$x_1 + z_{1.m+1}x_{m+1} + \ldots + z_{1N}x_N = x_1^0,$$
$$\vdots$$
$$x_m + z_{m.m+1}x_{m+1} + \ldots + z_{mN}x_N = x_m^0,$$

and the objective function C can be written as

$$C + z_{0.m+1}x_{m+1} + \ldots + z_{0N}x_N = z_{00}.$$

We shall have to deal very often with such relationships and write them as follows:

	x_{m+1}		x_N	
x_1	$z_{1.m+1}$		$z_{1.N}$	x_1^0
.				
.				
.				
x_m	$z_{m.m+1}$		$z_{m.N}$	x_m^0
C	$z_{0.m+1}$		$z_{0.N}$	z_{00}

This *tableau,* as we shall call it, describes the relationship clearly and briefly. The rows are allocated to the basic, and the columns to the nonbasic variables. The objective function has a further row allocated to it. The last column gives the current values of the basic variables, and of the objective function, while the other columns contain the coefficients of the nonbasic variables.

It will be convenient to exhibit alongside and above the variables their coefficient in the original expression of the objective function. Thus

$$
\begin{array}{ccccc}
 & c_{m+1} & & c_N & \\
 & x_{m+1} & & x_N & \\
c_1 & x_1 & \cdot\ \ \cdot\ \cdot & & x_1^0 \\
\cdot & & & & \\
\cdot & & & & \\
\cdot & & & & \\
c_m & x_m & \cdot\ \ \cdot\ \cdot & & x_m^0 \\
\hline
 & C & \cdot\ \ \cdot\ \cdot & & z_{00}
\end{array}
$$

We call this type of tableau *contracted,* to distinguish it from the *extended* tableau, which contains columns for the basic variables as well:

	x_1	x_2	x_m	x_{m+1}	x_N	
x_1	1	0	0	z_{1m+1}	z_{1N}	x_1^0
x_2	0	1	0	z_{2m+1}	z_{2N}	x_2^0
x_m	0	0	1	z_{mm+1}	z_{mN}	x_m^0
C	0	0	0	z_{0m+1}	z_{0N}	z_{00}

The successive steps of the Simplex method can be described in terms of manipulation of either type of tableau. In this method we move from one basic solution to an adjacent one. With a contracted tableau, when we exchange the hitherto basic variable x_r for the hitherto nonbasic x_h, and make the first nonbasic and the second basic, then we use the row of x_r which becomes now free, to enter the values belonging to x_h, and the column of x_h will be used for the values belonging to x_r. The order of subscripts in the rows as well as in the columns will thereby be permuted, but this is of no account in comparison with the resulting compactness of the presentation. This reflects also the manner in which one would allocate addresses in an electronic computer.

In the extended tableau the order of the columns remains fixed throughout. The columns containing just one single 1, and having in all other positions a 0, are those referring to the current basic variables. There need never be two such columns with the value 1 in the same row because, if this were so, then the sum of the two corresponding variables could be introduced instead of those two as a new single variable.

The rules of transforming tableaux have been derived in Chapter 2. We repeat them here, in words which refer to the restricted tableau. Let the basic variable x_r become nonbasic, and the nonbasic variable x_h become basic. Call z_{rh} the *pivot*, the row z_{rj} (j being the subscripts of nonbasic variables) the *pivotal row*, and the column z_{ih} (i being the subscripts of basic variables) the *pivotal column*. Then

(1) replace the pivot by its reciprocal,

(2) divide all remaining entries of the pivotal row by the pivot,

(3) divide all remaining entries of the pivotal column by the pivot and change the sign, and

(4) the remaining terms, z_{st} say, are replaced by $z_{st} - z_{rt} z_{sh}/z_{rh}$.

We illustrate now how the Simplex method works.

EXAMPLE 5–1. (Compare with Example 2–12.) Maximize $B = 3x_1 + 6x_2 + 2x_3$ subject to

$$3x_1 + 4x_2 + x_3 \leqslant 2,$$
$$x_1 + 3x_2 + 2x_3 \leqslant 1,$$
$$x_1, x_2, x_3 \geqslant 0,$$

or, equivalently, subject to

$$3x_1 + 4x_2 + x_3 + x_4 = 2,$$
$$x_1 + 3x_2 + 2x_3 + x_5 = 1,$$
$$x_1, x_2, x_3, x_4, x_5 \geqslant 0.$$

We start off with some basic feasible solution. In this case such a solution is immediately at hand:

$$x_4 = 2, \qquad x_5 = 1, \qquad x_1 = x_2 = x_3 = 0.$$

Hence x_4 and x_5 are basic, and the other variables are nonbasic. We write our first tableau as

		3	6	2	
		x_1	x_2	x_3	
0	x_4	3	4	1	2
0	x_5	1	3	2	1
B		-3	-6	-2	0

The bottom row indicates that

$$B - 3x_1 - 6x_2 - 2x_3 = 0.$$

The objective function B is expressed in terms of the nonbasic variables. Their current value is zero, and hence $B = 0$ at this stage. We must now determine whether it could be increased without contradicting any of the constraints. This depends on whether a change in the nonbasic variables would increase B. Such a change could only be an increase, because values smaller than zero are not allowed. Since all coefficients in the expression $B = 3x_1 + 6x_2 + 2x_3$ are positive (or, in other words, the coefficients in all columns of nonbasic variables in the B row of the tableau are negative), any of the variables could be increased with profit.

It is natural to choose the variable with the highest coefficient, i.e. that with the most negative entry in the bottom row of the tableau. In this case it is x_2. This choice is a convention, though, not a necessity (see Exercise 5–5).

We want to consider only basic solutions. Therefore, if we make x_2 positive, some other variable will have to be made zero. Which one?

This question is related to the question of how far x_2 can be increased. The only reason which sets a limit to its increase is the possibility that some other variable could become negative. Now if we write our constraints as

$$x_4 = 2 - 3x_1 - 4x_2 - x_3,$$
$$x_5 = 1 - x_1 - 3x_2 - 2x_3,$$

then we see that x_4 would become negative if x_2 were increased beyond $2/4$ (the other nonbasic variables, viz. x_1 and x_3 remain at zero), and x_5 would become negative if x_2 were increased beyond $1/3$. Because the latter ratio is smaller, it is the equation for x_5 which sets the limit. As the value of x_2 reaches $1/3$, x_5 becomes zero, and nonbasic. We have thus reached another basic feasible solution. If, by mistake, we had made an incorrect exchange of variables, then this would show by the new basic solution not being feasible, because it would then contain a negative variable.

Notice also that if the coefficient of x_2 had been positive in the equation for some basic variable, then no increase of x_2 would have reduced the latter. It follows, then, that:

(i) The variable to be made nonbasic is that which first becomes zero through an increase of the nonbasic variable chosen to be increased, say x_h. In terms of the tableau, it is determined by finding the smallest ratio x_i^0/z_{ih}, where i takes all values which are subscripts of basic variables, such that $z_{ih} > 0$.

(ii) If all coefficients z_{it} for some nonbasic variable x_t with negative z_{0t} are negative, then the maximizing problem has no finite solution.

In the present case we have decided to exchange x_5 for x_2. The rules of tableau transformation then lead to

		3		2	
		x_1	x_5	x_3	
	x_4	5/3*	—4/3	—5/3	2/3
6	x_2	1/3	1/3	2/3	1/3
	B	—1	2	2	2

Now the coefficient z_{01} is negative, and $2/3:5/3$ is smaller than $1/3:1/3$, hence we exchange x_4 and x_1. The asterisk in the tableau indicates the pivot.

				2	
		x_4	x_5	x_3	
3	x_1	3/5	—4/5	—1	2/5
6	x_2	—1/5	3/5	1	1/5
	B	3/5	6/5	1	12/5

This is the final tableau.

The relations (2–9) and (2–8) provide a rough check on the entries in a tableau. For instance, taking the first constraint

$$3 \times \tfrac{2}{5} + 4 \times \tfrac{1}{5} = 2 \quad \text{(Right-hand side of first constraint)},$$

$$3 \times \tfrac{3}{5} + 4 \times -\tfrac{1}{5} = 1 \quad \text{(Coefficient of } x_4 \text{ in first constraint)},$$

$$3 \times -\tfrac{4}{5} + 4 \times \tfrac{3}{5} = 0 \quad \text{(Coefficient of } x_5 \text{ in first constraint)},$$

$$3 \times -1 + 4 \times 1 = 1 \quad \text{(Coefficient of } x_3 \text{ in first constraint)}.$$

Similarly, from the second constraint

$$\tfrac{2}{5} + 3 \times \tfrac{1}{5} = 1,$$

$$\tfrac{3}{5} + 3 \times -\tfrac{1}{5} = 0,$$

$$-\tfrac{4}{5} + 3 \times \tfrac{3}{5} = 1,$$

$$-1 + 3 \times 1 = 2.$$

Relations (2–c) and (2–d) tell us that we must have

$$3 \times \tfrac{2}{3} + 6 \times \tfrac{1}{3} = \tfrac{12}{3},$$
$$3 \times \tfrac{3}{3} + 6 \times -\tfrac{1}{3} = \tfrac{3}{3},$$
$$3 \times -\tfrac{4}{3} + 6 \times \tfrac{3}{3} = \tfrac{6}{3},$$
$$3 \times -1 + 6 \times 1 - 2 = 1.$$

Of course, these are not very comprehensive checks, because any number multiplied by zero gives the right answer even if the number itself is incorrect. For this reason it is preferable to introduce a fictitious objective function, none of whose coefficients is zero, in an additional row. Thus if we take as a fictitious objective function $\sum_{i=1}^{N} x_i$, the entry in the x_j column of the check row will be $\sum_i z_{ij} + z_{0j} - 1$, where the summation extends over the subscripts of all basic variables. An analogous entry will be made in the last column (that of the values of the variables). The relationship expressed in the last formula remains valid after transformation as well, and this supplies a convenient check at each stage.

In Example 5–1 this means that

	x_1	x_2	x_3			x_1	x_2	x_3	
x_4	3	4	1	2	x_4	5/3	−4/3	−5/3	2/3
x_5	1	3*	2	1	x_2	1/3	1/3	2/3	1/3
B	−3	−6	−2	0	B	−1	2	2	2
Check	0	0	0	2	Check	0	0	0	2

and so on.

5–2 Simplex method. Finding a first feasible solution. The next example will show how we may try to find a first feasible solution, where none is immediately apparent.

EXAMPLE 5–2. Given

$$3x_1 + \ \ x_2 - x_3 = 3,$$
$$4x_1 + 3x_2 - x_4 = 6, \qquad x_1, x_2, x_3, x_4, x_5 \geqslant 0.$$
$$x_1 + 2x_2 - x_5 = 2,$$

$$\text{Minimize } 2x_1 + x_2 = C.$$

We may imagine that the constraints originated from inequalities, and that x_3, x_4, and x_5 are additional, or slack, variables.

In this case it is not easy to find a feasible solution of the constraints. We must expect such cases to arise, because we know that constraints can be inconsistent, and in that case no feasible solution exists at all.

Here we make use of the device of introducing *artificial variables*. We modify our problem so as to obtain one which has a feasible solution that is immediately apparent. Thus

$$3x_1 + x_2 - x_3 + x_{101} = 3,$$
$$4x_1 + 3x_2 - x_4 + x_{102} = 6, \qquad x_1, \ldots, x_5,$$
$$x_1 + 2x_2 - x_5 + x_{103} = 2, \qquad x_{101}, x_{102}, x_{103} \geqslant 0.$$

The variables x_{101}, x_{102}, and x_{103} are called artificial. A feasible solution of these constraints is $x_{101} = 3$, $x_{102} = 6$, $x_{103} = 2$.

Of course this is not the problem we really want to solve. But it does not matter if the artificial variables appear in some solutions on the way to the optimum, provided we do not find them with a nonzero value in the final solution.

Let us, therefore, attach to the artificial variables a very high penalty. Instead of C above, we introduce

$$C_M = 2x_1 + x_2 + M(x_{101} + x_{102} + x_{103}),$$

where M is larger than any number with which it is ever compared during the subsequent computations. If there is an answer at all to the original problem, then there is an answer to the modified problem with all artificial variables zero. For such a set of variables will make C_M smaller than any set in which a positive artificial variable appears. Hence, if there is an answer at all to the original problem, then we shall find it by solving the modified one, and no artificial variable will be positive in the final answer. On the other hand, if the optimal solution to the modified problem contains an artificial variable with a positive value, then the original problem has inconsistent constraints, and no answer exists.

A similar argument shows that if we wish to maximize an objective function, then artificial variables should be introduced with factors $-M$.

It is not always necessary to introduce an artificial variable into every constraint. For instance, if the constant on the right-hand side of the third constraint had been -2, then x_{103} would be unnecessary, and $x_5 = 2$ could be a basic variable to begin with.

Whenever an artificial variable becomes nonbasic, and hence zero, then we can omit this variable from further consideration. We cannot omit it, however, if it has the value zero as a basic variable in a degenerate case because, if we did, then we should be left with too few basic variables.

We shall now solve Example 5–2 in a succession of tableaux.

			2	1				
		x_1	x_2	x_3	x_4	x_5		
M	x_{101}	3*	1	−1			3	
M	x_{102}	4	3		−1		6	
M	x_{103}	1	2			−1	2	
C	$\begin{cases} \\ \\ \end{cases}$	−2	−1					
		8	6	−1	−1	−1	11	

The expression for C is distributed over two rows, and it is under-stood that the second contains the coefficients of M and they are there-fore overriding when compared with those in the penultimate row. The last row is constructed by adding all the coefficients z_{ij} in the same column.

We choose x_1 to be made basic, because $8M - 2$ is positive, and larger than $6M - 1$. Of the ratios 3/3, 6/4, 2/1, the first is smallest. Therefore we exchange x_1 and x_{101}. The latter becomes basic and can henceforth be ignored.

			1				
		x_2	x_3	x_4	x_5		
2	x_1	1/3	−1/3			1	
M	x_{102}	5/3	4/3	−1		2	
M	x_{103}	5/3*	1/3		−1	1	
C	$\begin{cases} \\ \\ \end{cases}$	−1/3	−2/3			2	
		10/3	5/3	−1	−1	3	

			x_3	x_4	x_5		
2	x_1		−2/5		1/5	4/5	
M	x_{102}		1	−1	1*	1	
1	x_2		1/5		−3/5	3/5	
C	$\begin{cases} \\ \\ \end{cases}$	−3/5		−1/5	11/5		
		1	−1	1	1		

We could now make either x_3 or x_5 basic. If we want to be subtle, then we take that variable whose exchange reduces C to a larger extent.

The row connected with M will disappear in any case, and $M + 11/5$ will be decreased by $M - 3/5$ if we choose x_3, or by $M - 1/5$ if we choose x_5. Hence we exchange x_5 and x_{102}.

		x_3	x_4	
2	x_1	$-3/5$	$1/5$	$3/5$
	x_5	1	-1	1
1	x_2	$4/5$	$-3/5$	$6/5$
	C	$-2/5$	$-1/5$	$12/5$

This is, then, the answer:

$$x_1 = 3/5, \quad x_2 = 6/5, \quad x_3 = 0, \quad x_4 = 0, \quad x_5 = 1, \quad C = 12/5.$$

If we carry the artificial variables along after they have become nonbasic, then the coefficient of C in their column would remain at $-M$.

As a matter of fact, the method which we have described is unnecessarily clumsy. We can obtain a first feasible solution by introducing just one single artificial variable, x_0 say, as follows:

$$3x_1 + x_2 - x_3 + x_0 = 3,$$
$$4x_1 + 3x_2 - x_4 + x_0 = 6,$$
$$x_1 + 2x_2 - x_5 + x_0 = 2.$$
$$\text{Minimize } 2x_1 + x_2 + Mx_0.$$

The largest positive value on the right is 6, and we make $x_0 = 6$, and then $x_3 = 6 - 3 = 3$, and $x_5 = 6 - 2 = 4$. Thus we have again three basic variables. Solving for these, we obtain the following tableau:

			2	1	
		x_1	x_2	x_4	
M	x_0	4	3	-1	6
	x_3	1	2	-1	3
	x_5	3^*	1	-1	4
	C $\begin{cases} \\ \\ \end{cases}$	$\begin{matrix} -2 \\ 4 \end{matrix}$	$\begin{matrix} -1 \\ 3 \end{matrix}$	-1	6

The reader will not have any difficulty in continuing. The final answer will, of course, be the same as before.

The *M-method* is not the only one used to obtain a first feasible solution. Another possibility, the *Two-phase method* (cf. [55], [120]), will now be explained. Let it be required to minimize

$$C = c_1 x_1 + \ldots + c_N x_N,$$

subject to

$$\sum_{=1}^{N} a_{ij} x_i = b_j \; (\geqslant 0) \qquad (j = 1, \ldots, m)$$

and

$$x_i \geqslant 0 \qquad (i = 1, \ldots, N).$$

We add another equation, linearly dependent on the above and thus not adding anything new:

$$-\sum_{i=1}^{N} \sum_{j=1}^{m} a_{ij} x_i = -\sum_{j=1}^{m} b_j.$$

We also rewrite the constraints with an artificial variable in each of them:

$$a_{11} x_1 + \ldots + a_{N1} x_N + x_{N+1} \quad = b_1,$$

.

.

.

$$a_{1m} x_1 + \ldots + a_{Nm} x_N + x_{N+m} \quad = b_m,$$
$$-\sum_j a_{1j} x_1 - \ldots - \sum_j a_{Nj} x_N - x_{N+m+1} = -\sum_j b_j.$$

The variables $x_{N+1}, \ldots, x_{N+m}$ are to be non-negative.

If we add all these equations, we find that

$$x_{N+1} + \ldots + x_{N+m} = x_{N+m+1},$$

so that the latter variable will also have only non-negative values. We add, also, the equation

$$-C + c_1 x_1 + \ldots + c_N x_N = 0,$$

with C as a variable without sign restriction, and tackle first the problem of minimizing x_{N+m+1}.

The minimum will be zero if and only if all other artificial variables have the value zero. Hence, in this case, we have obtained a solution to the original problem as well, and then proceed by the Simplex method, if the solution is not yet optimal.

EXAMPLE 5-3. Write our last example as follows:

$$3x_1 + x_2 - x_3 + x_{N+1} = 3,$$
$$4x_1 + 3x_2 - x_4 + x_{N+2} = 6,$$
$$x_1 + 2x_2 - x_5 + x_{N+3} = 2,$$
$$2x_1 + x_2 - C = 0,$$
$$-8x_1 - 6x_2 + x_3 + x_4 + x_5 - x_{200} = -11.$$

Minimize x_{200}, say.

If the reader carries out the computations, he will find that the successive tableaux are essentially the same as those in the M-method. The fourth tableau will be

	x_{N+1}	x_{N+3}	x_3	x_4	x_{N+2}	
x_1	3/5	0	-3/5	1/5	-1/5	3/5
x_5	-1	-1	1	-1	1	1
x_2	-4/5	0	4/5	-3/5	3/5	6/5
C	2/5	0	-2/5	-1/5	1/5	12/5
x_{200}	-1	-1	0	0	-1	0

At this stage we discard the artificial variables and proceed with the remaining tableau. As it happens, it already gives the final answer. However, if we wished to maximize the objective function, then the first phase would have been the same as in the present case, and the second phase would have to be entered.

It may happen at the termination of the first phase, i.e. when the variable x_{N+m+1} is zero, that there remains an artificial variable in the basis, though of course with value zero. It would be desirable to make such a variable nonbasic, because it could afterwards be ignored altogether.

Let x_h be the nonbasic variable to be made basic. Let the artificial variable which we should like to make nonbasic be x_s. If z_{sh} is positive, then x_s qualifies for removal from the basic variables according to the ordinary rules. If z_{sh} is zero, then it does not qualify, and hence the exchange cannot affect x_s, which remains basic, but still with value zero. However, if $z_{sh} < 0$, then the exchange of x_h and some variable other than x_s, say x_r, would change $x_s^0 = 0$ into $x_s^0 - z_{sh}x_r^0/z_{rh}$ which might be positive. To avoid this, we decide that z_{sh}, even if it is negative, should serve as a pivot. Then x_s will become nonbasic, x_h will become basic (with value zero), and all other basic variables retain their values. (This suggestion is due to G. F. Hadley and M. A. Simmonard [78].)

If at any stage we should find $z_{sj} = 0$ for all j and for all s corresponding to artificial variables, then the matrix of coefficients of the original problem has rank less than m and therefore some constraint must be redundant. In such a case the artificial variable would have to remain basic to the end, because without it it would be impossible to find m basic variables. (See Exercise 5–8.)

Yet another method is mentioned in [115]. Suppose that $\sum_{i=1}^{N} c_i x_i$ is to be minimized subject to

$$\sum_{i=1}^{N} a_{ij} x_i = b_j, \qquad x_i \geqslant 0.$$

We set

$$x_{m+1} = \ldots = x_N = 0,$$

and solve for $x_1, \ldots, x_m$. Let the solution be

$$x_1^0, \ldots, x_k^0 < 0, \qquad x_{k+1}^0, \ldots, x_m^0 \geqslant 0, \qquad x_{m+1}^0 = \ldots = x_N^0 = 0.$$

Then

$$x_1 = \ldots = x_k = 0, \qquad x_{k+1} = x_{k+1}^0, \ldots, x_N = x_N^0,$$
$$y_1 = -x_1^0, \ldots, y_k = -x_k^0$$

is a basic feasible solution of

$$\sum_{i=1}^{N} a_{ij} x_i + \sum_{i=1}^{k} (-a_{ij}) y_i = b_j, \qquad x_i, y_i \geqslant 0,$$

and the problem of minimizing

$$\sum_{i=1}^{N} c_i x_i + M \sum_{i=1}^{k} y_i \qquad (M \text{ very large})$$

subject to the latter constraints has the same optimal solution as the original problem.

For convenience, we survey now the elementary rules of the Simplex method:

Transform all constraints into equations, by adding additional variables where necessary.

Find a basic feasible solution, by adding artificial variables where necessary.

Construct the tableau, using appropriate checks.

If, when minimizing (or maximizing), there is a positive (or a negative) value amongst the z_{0j}, choose one of the corresponding variables to make it basic. Let this be x_h.

Choose the variable x_r, to exchange for x_h, by determining the smallest ratio x_i^0 / z_{ih}, considering only those basic x_i for which z_{ih} is positive. (If there is no such x_i, then the problem has no bounded solution.)

Apply the rules of tableau transformation.

Whenever an artificial variable becomes nonbasic, ignore it afterwards.

If all the z_{0j} are positive when maximizing, or negative when minimizing, then the optimum is reached.

If at this stage there is still an artificial variable with positive value in the basis, then the original problem has no solution.

We add the following remark: Whenever a z_{0j} is zero, then the corresponding nonbasic variable could be exchanged. The rules of tableau transformation show that then the objective function would not change its value, but another set of basic variables would emerge. Any linear combination of the two solutions gives the same value to the objective function, but would not be basic.

The fact that such an exchange cannot affect the value of the objective function is also clear from first principles: if z_{0j} is zero, then the variable x_j does not appear in the expression of the objective function, in terms of nonbasic variables, and hence an increase of x_j does not affect the value of the objective function.

At each tableau transformation, when x_r is exchanged for x_h, the value of the objective function is reduced by $x_r^0 z_{0h}/z_{rh}$. The denominator, i.e. the pivot, is positive, and if the value x_r^0 of the variable x_r is also positive, then the change in the value of the objective function will be positive when z_{0h} is negative, and vice versa. Hence we shall obtain an increase in a maximizing problem and a decrease in a minimizing problem, as it ought to be. Therefore, always assuming that x_r^0 is positive (not zero), no set of basic variables previously encountered can reappear because that would imply also a reappearance of a previous value of the objective function. (Individual variables might reappear in the basis, though.)

Since there exists only a finite number of sets of basic variables $\left[\text{namely} \begin{pmatrix} N \\ m \end{pmatrix} \right]$, the opt mum must be reached after a finite number of steps.

5–3 Simplex method. Degeneracy. This argument breaks down if at any time some basic x_r is zero. It is this degenerate case which will occupy us now.

EXAMPLE 5–4. Given

$$3x_1 + x_2 - x_3 = 3,$$
$$4x_1 + 3x_2 - x_4 = 5,$$
$$x_1 + 2x_2 - x_5 = 2.$$

Minimize $2x_1 + x_2 = C$, say.

Using the version of the M-method which uses just one artificial variable, we have

		2	1		
		x_1	x_2	x_4	
M	x_0	4	3	-1	5
	x_3	1	2	-1	2
	x_5	3*	1	-1	3
C		-2	-1	0	0
		4	3	-1	5

			1		
		x_5	x_2	x_4	
M	x_0	$-4/3$	$5/3$*	$1/3$	1
	x_3	$-1/3$	$5/3$	$-2/3$	1
2	x_1	$1/3$	$1/3$	$-1/3$	1
C		$2/3$	$-1/3$	$-2/3$	2
		$-4/3$	$5/3$	$1/3$	1

Now x_2 should be made basic. The ratio x_i^0/z_{ih} is the same for x_0 and for x_3, viz. $3/5$. This is an indication of degeneracy, as has been shown in Section 2–1.

We can choose either variable to be nonbasic, without thereby forcing any variable to become negative. We choose x_0 (since this is a variable we want to remove in any case):

		x_5	x_4	
1	x_2	$-4/5$	$1/5$	$3/5$
	x_3	1*	-1	0
2	x_1	$3/5$	$-2/5$	$4/5$
C		$2/5$	$-3/5$	$11/5$

The degeneracy has become manifest. The variable to be made basic is now x_5, the pivot is 1, and the next tableau is

		x_3	x_4	
1	x_2	$4/5$	$-3/5$	$3/5$
	x_5	1	-1	0
2	x_1	$-3/5$	$1/5$	$4/5$
C		$-2/5$	$-1/5$	$11/5$

The values of the variables have not changed, even though the division into basic and nonbasic ones has. The value of C has also remained unaltered, and this is precisely the point that made us study such a case. If we cannot reduce the value of C at each step, then we cannot be confident that we shall ever reach the final answer.

In this particular example it happens that we have already reached the minimum. But clearly, this need not be so in all cases. The first question to ask is therefore this: Is it conceivable that in a degenerate case we go through a *cycle*, in the sense that the value of the objective function remains unaltered through a number of steps, that we return to a set of basic solutions which we have encountered before, and that from then on we pass through an unending repetition of the same succession of steps?

The answer to this question depends on the rule, or lack of rule, for resolving a tie amongst the ratios which decide what variable to make nonbasic. If we adopt an unsuitable rule, then we may very well find ourselves in a loop, as the following example (constructed by E. M. L. Beale, see [25]) shows:

	3/4	−20	1/2	−6	
	x_1	x_2	x_3	x_4	
x_5	1/4	−8	−1	9	0
x_6	1/2	−12	−1/2	3	0
x_7	0	0	1	0	1
B	−3/4	20	−1/2	6	0

If we make it our rule to choose, whenever there is a tie, the basic variable with the lower subscript to become nonbasic, and make that variable basic whose z_{0j} is most negative, then we get into a cycle. We leave the computation to the reader, but indicate here the succession of basic variables; their subscripts are

$(5, 6, 7)$, $(1, 6, 7)$, $(1, 2, 7)$, $(3, 2, 7)$, $(3, 4, 7)$, $(5, 4, 7)$, $(5, 6, 7)$.

We have returned to the set with which we started.

This example was deliberately constructed to show that cycles could occur. Another unpublished but earlier example is due to A. J. Hoffman. In practice cycles do not appear. It could therefore be argued that we need not worry about them, and many programmes for automatic computation ignore the possibility. But from a theoretical point of view it is of interest to have a rule of choice, whenever a tie appears, which can be proved never to lead to a cycle.

The idea behind such a rule will best be appreciated if we consider a geometrical representation. We have seen in Chapter 2 that a degenerate case arises, for $N = m + 2$, when more than two lines representing $x_i = 0$ pass through the same point. In general, degeneracy means that more than $N - m = n$ hyperplanes pass through the same vertex.

A natural way of dealing with such a degenerate case is to apply a *perturbation* to the hyperplanes, in such a way that they move off and that not more than $N - m$ of them pass through any one point. We imagine each hyperplane to remain parallel to its original position, and the displacement to be small enough so that they do not pass across any other vertex of the feasible region. For instance, if we have the configuration (a) of Fig. 5–1 we might change it into (b) but not into (c).

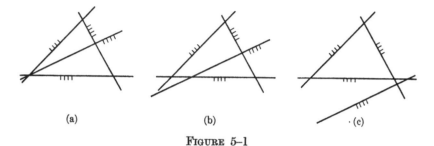

(a)   (b)   · (c)

FIGURE 5–1

EXAMPLE 5–5. Given

$$x_1 - 2x_2 + x_3 = 0,$$
$$2x_1 - 3x_2 + x_4 = 0,$$
$$x_1 + 4x_2 + x_5 = 1.$$

Four lines intersect in the origin, viz. $x_1 = 0$, $x_2 = 0$, $x_3 = 0$, $x_4 = 0$. This point can therefore be defined in $\binom{4}{2} = 6$ different ways as the intersection of two lines. Here we have six tableaux which describe the same point, each with another set of two nonbasic variables. (No objective function is given, since it is irrelevant for the present argument.)

a	x_1	x_2		b	x_1	x_3		c	x_1	x_4	
x_3	1	-2	0	x_2	$-1/2$	$-1/2$	0	x_2	$-2/3$	$-1/3$	0
x_4	2	-3	0	x_4	$1/2$	$-3/2$	0	x_3	$-1/3$	$-2/3$	0
x_5	1	4	1	x_5	3	2	1	x_5	$11/3$	$4/3$	1

d					e					f		
	x_2	x_3				x_2	x_4				x_3	x_4
x_1	-2	1	0		x_1	$-3/2$	$1/2$	0		x_1	-3	2	0
x_4	1	-2	0		x_3	$-1/2$	$-1/2$	0		x_2	-2	1	0
x_5	6	-1	1		x_5	$11/2$	$-1/2$	1		x_5	11	-6	1

If we now perturb the straight lines through the origin, then we
obtain one of the configurations of Fig. 5–2.

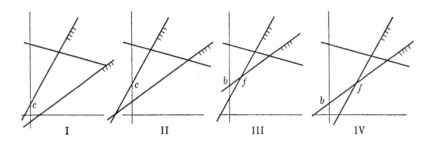

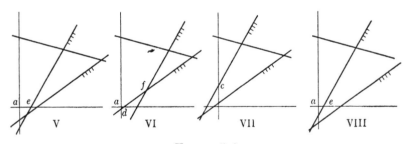

FIGURE 5–2

The intersections of the perturbed lines have been marked with the
same letters as in their description in tableau form. The essential
feature to be observed is the combination of pairs which are now
adjacent. For instance, a is adjacent to e in V or VIII, and to d in VI.

As we know, in the Simplex method we move from a vertex to an
adjacent one, in a direction in which the arrow indicating the preferred
direction has a positive component. Once we have decided on a
specific perturbation, cycling is impossible, because having started
from some point we could only return to it through a succession of

vectors adding up to zero, and therefore they could not possibly all have positive components in the preferred direction. Thus cycling can only occur when we change the perturbed configuration during the procedure, so that having moved from one point to an adjacent one in one configuration, we then find the corresponding one in another, and move on from there.

No cycling is possible in this example. The simplest cycle known occurs in the case $N = m + 4$; it is that shown above.

It is possible to give a rule which is certain to avoid cycles in any case. In geometrical terms, it means sticking to one single perturbation throughout. We explain here the rule introduced by A. Charnes in [34].

Consider a basic solution with basic variables $x_1, \ldots, x_m$, represented by the following system of equations:

$$x_i + z_{i,m+1}x_{m+1} + \ldots + z_{iN}x_N = x_i^0 \qquad (i = 1, \ldots, m).$$

We now change the right-hand side by adding to it a certain polynomial in ε, where ε is a small positive fraction, so small that any multiple of it which appears in the process is still smaller than any other number. Of course, higher powers of ε are smaller than lower ones. The polynomial that we add to the right-hand side in the ith row is

$$\varepsilon^i + z_{i,m+1}\varepsilon^{m+1} + \ldots + z_{iN}\varepsilon^N.$$

No tie is then possible. For either $x_k^0/z_{ks} \neq x_l^0/z_{ls}$, and then no combination of powers of ε can make the two ratios equal, or $x_k^0/z_{ks} = x_l^0/z_{ls}$, and then $\varepsilon^k/z_{ks} \neq \varepsilon^l z_{ls}$, and again the other powers of ε cannot produce equality, since all powers $\varepsilon^{m+1}, \ldots, \varepsilon^N$ are different from either ε^k or ε^l.

It will be seen that in order to obtain this result, it would have been sufficient to add to the right-hand side ε^i, as suggested by G. B. Dantzig at an earlier time. With the polynomials as we have introduced them it is immediately clear that the coefficients of the various powers of ε are transformed in the same way as are the coefficients of the variables with the same subscripts as the exponents of ε. The coefficients of $\varepsilon, \ldots, \varepsilon^m$ are transformed in the same way as are the coefficients of $x_1, \ldots, x_m$, provided these are also shown in the tableau. In other words, it is recommended to think of an extended rather than of a contracted tableau.

After transformation we remember that any higher power of ε has a smaller value than any lower one. However, there is no point in writing all the coefficients of the polynomial in ε, because they are already there in the extended tableau.

In [34] it is also pointed out that the suggested method will never perturb away the whole feasible region, as might happen with other methods, if the feasible region is of smaller dimensionality than $N - m$.

EXAMPLE 5–6. We take the example on which we have illustrated the perturbations. For an objective function we take $B = x_1 + x_2$, to be maximized. We can then start with (perturbation VI)

(a)

	1	1				
	x_1	x_2	x_3	x_4	x_5	
x_3	1*	−2	1	0	0	$\varepsilon - 2\varepsilon^2 + \varepsilon^3$
x_4	2	−3	0	1	0	$2\varepsilon - 3\varepsilon^2 + \varepsilon^4$
x_5	1	4	0	0	1	$1 + \varepsilon + 4\varepsilon^2 + \varepsilon^5$
B	−1	−1	0	0	0	$-\varepsilon - \varepsilon^2$

The value of B is given as $-\varepsilon - \varepsilon^2$, so that the rule connecting the values of the last column with the coefficients in the other columns holds for B as well.

To find the variable which leaves the basis, we compare the ratios

$$(\varepsilon - 2\varepsilon^2 + \varepsilon^3) \div 1 \quad \text{and} \quad (2\varepsilon - 3\varepsilon^2 + \varepsilon^4) \div 2.$$

The first is smaller, because $\varepsilon - 2\varepsilon^2 < \varepsilon - \tfrac{3}{2}\varepsilon^2$, while the powers ε^3 and ε^4 are too small to affect this inequality, whatever their coefficients. Hence we reach

(d)

		1	1				
		x_1	x_2	x_3	x_4	x_5	
1	x_1	1	−2	1	0	0	$\varepsilon - 2\varepsilon^2 + \varepsilon^3$
	x_4	0	1*	−2	1	0	$\varepsilon^2 - 2\varepsilon^3 + \varepsilon^4$
	x_5	0	6	−1	0	1	$1 + 6\varepsilon^2 - \varepsilon^3 + \varepsilon^5$
	B	0	−3	1	0	0	$-3\varepsilon^2 + \varepsilon^3$

B has strictly increased. The values of the basic variables can be transformed according to the usual rules, but the coefficients of the various powers of ε can equally well be read off the other columns of the tableau:

(f)

		1	1				
		x_1	x_2	x_3	x_4	x_5	
1	x_1	1	0	-3	2	0	$\varepsilon - 3\varepsilon^3 + 2\varepsilon^4$
1	x_2	0	1	-2	1	0	$\varepsilon^2 - 2\varepsilon^3 + \varepsilon^4$
	x_5	0	0	11*	-3	1	$1 + 11\varepsilon^3 - 3\varepsilon^4 + \varepsilon^5$
	B	0	0	-5	3	0	$-5\varepsilon^3 + 3\varepsilon^4$

,

		1	1				
		x_1	x_2	x_3	x_4	x_5	
1	x_1	1	0	0	13/11	3/11	$\frac{3}{11} + \varepsilon + \frac{13}{11}\varepsilon^4 + \frac{3}{11}\varepsilon^5$
1	x_2	0	1	0	5/11	2/11	$\frac{2}{11} + \varepsilon^2 + \frac{5}{11}\varepsilon^4 + \frac{2}{11}\varepsilon^5$
	x_3	0	0	1	$-3/11$	1/11	$\frac{1}{11} + \varepsilon^3 - \frac{3}{11}\varepsilon^4 + \frac{1}{11}\varepsilon^5$
	B	0	0	0	18/11	5/11	$1 + \frac{18}{11}\varepsilon^4 + \frac{5}{11}\varepsilon^5$

.

The answer is $x_1 = 3/11$, $x_2 = 2/11$, $x_3 = 1/11$, $x_4 = 0$, $x_5 = 0$.

The point we want to make can now be introduced by studying tableau (a). Instead of writing the powers of ε, we look at the coefficients in the other columns. Thus, having found that $0/1 = 0/2$, which gives a tie, we carry on dividing the coefficients of the successive columns by the values in the pivotal column; we compare $-2/1$ and $-3/2$. If there were again a tie, then we would take column x_3, x_4, etc., considering the ties which still remain, until they are all broken. This must happen some time, because we are dealing with an extended tableau, and there is a column with only one single 1 in it, while the other entries are all 0.

It is easily seen that, in order to apply this rule, the order of subscripts of the variables in the extended tableau must remain unaltered throughout.

5–4 The inverse matrix method. The method which we are about to discuss is based on the Simplex method and deviates from it only in detail, for the conveniences of computation.

Consider, once more, an extended tableau where, of course, those columns referring to basic variables have a single 1, and otherwise only 0 in their columns.

In the Inverse Matrix method we use again only a selection of columns of the extended tableau, but not precisely those which constituted the contracted tableau for the Simplex method. In fact, this time we deal with the same selection of columns throughout.

We start with the equations

$$x_{n+j} + a_{1j}x_1 + \ldots + a_{nj}x_n = b_j \geqslant 0 \qquad (j = 1, \ldots, m).$$

$$\text{Minimize } c_1 x_1 + \ldots + c_n x_n.$$

Then the first solution is $x_{n+j} = b_j$ $(j = 1, \ldots, m)$, and the matrix M (cf. Section 2–3) is

$$M = \begin{pmatrix} b_1 & a_{11} & a_{n1} & 1 & 0 & 0 \\ \cdot & & & & & \\ \cdot & & & & & \\ \cdot & & & & & \\ b_m & a_{1m} & a_{nm} & 0 & 1 & 0 \\ 0 & -c_1 & -c_n & 0 & 0 & 1 \end{pmatrix}.$$

This contains, at the same time, the extended tableau for the first solution. $M_{n+1, \ldots, n+m}$ is a unit matrix, and so is its inverse $D_{n+1, \ldots, n+m}$.

Since we want to minimize, we have to look at the $-c_1, \ldots, -c_n$, to see whether any of these is positive. Let $-c_n$ be positive; then x_n qualifies for consideration to be made basic. It will be exchanged for that basic variable for which the ratio b_j/a_{nj} is smallest, considering only those a_{nj} which are positive. Say this happens for x_{n+1}, so that the next basic variables are

$$x_n, x_{n+2}, \ldots, x_{n+m}.$$

We could now compute the new extended tableau. But it is of the essence of the Inverse Matrix method that we make a selection of the columns and retain only those that make up $M_{n+1, \ldots, n+m}$. Only these will be transformed. In the present case we have

	x_{n+1}	x_{n+2}	x_{n+m}	
x_n	$1/a_{n1}$	0	0	0
x_{n+2}	$-a_{n2}/a_{n1}$	1	0	0
x_{n+m}	$-a_{nm}/a_{n1}$	0	1	0
	c_n/a_{n1}	0	0	1

and, in general,

$$\begin{pmatrix} z_{v_1 n+1} & z_{u_1 n+m} & 0 \\ z_{u_2 n+1} & z_{u_2 n+m} & 0 \\ \cdot & & \\ \cdot & & \\ \cdot & & \\ z_{u_m n+1} & z_{u_m n+m} & 0 \\ z_{0,n+1} & z_{0,n+m} & 1 \end{pmatrix}.$$

This selection from the extended tableau, together with the very first tableau M is sufficient to give all the information necessary to carry out the further steps.

To prove this, we mention that the matrix just written is the inverse of $M_{u_1, \ldots, u_m}$, i.e. it is identical with

$$D_{u_1, \ldots, u_m} = \begin{pmatrix} d_{1,u_1} & d_{m,u_1} & 0 \\ \cdot & & \\ \cdot & & \\ \cdot & & \\ d_{1,u_m} & d_{m,u_m} & 0 \\ \sum_s c_{u_s} d_{1u_s} & \sum_s c_{u_s} d_{mu_s} & 1 \end{pmatrix}.$$

This follows from (2–8) in Section 2–3.

To see what step to take next, we need certain entries of the extended tableau. But we know (see Chapter 2) that the latter is essentially $D_{u_1, \ldots, u_m} \times M$, so that all the information contained in the extended tableau can be retrieved by carrying out computations implied in this product. However, since we need only some of the entries, only some of the multiplications will have to be carried out.

We shall certainly want to know the values of the basic variables. To obtain them, we form the inner product of the several rows of $D_{u_1, \ldots, u_m}$ and the first column of M. Thus

$$d_{1u_i} b_1 + \ldots + d_{mu_i} b_m = x_{u_i}^0 \qquad \text{[cf. (2–a)].}$$

The inner products of the last row of $D_{u_1, \ldots, u_m}$ and the several columns of M (after the first) form the last row of the tableau. Thus

$$a_{i1} \sum_s c_{u_s} d_{1u_s} + \ldots + a_{im} \sum_s c_{u_s} d_{mu_s} - c_i = z_{0i} \qquad \text{[cf. (2–b) and (2–d)].}$$

Having decided on the variable which ought to become basic, x_h say, we need the values in the x_h column of the tableau. They are the inner products of the rows of $D_{u_1, \ldots, u_m}$ and the x_h column in M. Thus

$$d_{1u_j}a_{h1} + \ldots + d_{mu_j}a_{hm} = z_{u_jh} \quad [\text{cf. (2–b)}].$$

When the variables to be exchanged have been determined, say x_h for x_{u_r}, then the matrix $D_{u_1, \ldots, u_m}$ must be transformed. Because this is a square matrix, the transformation can be described in a simplified manner. Consider $D_{u_1, \ldots, u_m}$ and the x_h column

$$
\begin{matrix}
z_{u_1h} \\
\cdot \\
\cdot \\
\cdot \\
z_{u_mh} \\
z_{0h}.
\end{matrix}
$$

The latter may, actually, be one of the columns in the former. The matrix as well as the column are, in any case, parts of the extended tableau of the stage considered. Therefore $D_{u_1, \ldots, u_m}$ is transformed according to the familiar rules, and the result can be written as the product of two matrices, as follows:

$$
\begin{pmatrix}
1 & 0 & \ldots & -z_{u_1h}/z_{u_rh} & \ldots & 0 & 0 \\
\cdot & & & & & & \\
\cdot & & & & & & \\
\cdot & & & & & & \\
0 & 0 & \ldots & 1/z_{u_rh} & \ldots & 0 & 0 \\
\cdot & & & & & & \\
\cdot & & & & & & \\
\cdot & & & & & & \\
0 & 0 & \ldots & -z_{u_mh}/z_{u_rh} & \ldots & 1 & 0 \\
0 & 0 & \ldots & -z_{0h}/z_{u_rh} & \ldots & 0 & 1
\end{pmatrix}
\times D_{u_1, \ldots, u_m}.
$$

The constants in the first matrix are already known at this stage.

The successive inverse matrices arise from the first (which will often be the unit matrix) through premultiplications with matrices of the type shown. In automatic computation only one column need be recorded. (See [54].)

Since $M_{u_1, \ldots, u_m}$ was not singular, $D_{u_1, \ldots, u_m}$ is not singular either, and this remains true after all transformations.

The Inverse Matrix method is sometimes referred to as the Revised Simplex method. Its specific computational advantages as compared with the Simplex method are investigated by Harvey M. Wagner in [121].

EXAMPLE 5-7. (Same as Example 5-1.) Given

$$3x_1 + 4x_2 + x_3 \leqslant 2,$$
$$x_1 + 3x_2 + 2x_3 \leqslant 1.$$

Maximize $3x_1 + 6x_2 + 2x_3$.

We have

$$M = \begin{matrix} & x_1 & x_2 & x_3 & x_4 & x_5 & & \\ & \begin{pmatrix} 2 & 3 & 4 & 1 & 1 & 0 & 0 \\ 1 & 1 & 3 & 2 & 0 & 1 & 0 \\ 0 & -3 & -6 & -2 & 0 & 0 & 1 \end{pmatrix} & \begin{matrix} x_4 \\ x_5 \\ \end{matrix} \end{matrix}$$

and

$$D_{4,5} = \begin{pmatrix} 1 & 0 & 0 \\ 0 & 1 & 0 \\ 0 & 0 & 1 \end{pmatrix}.$$

We want to make x_2 basic and x_5 nonbasic. This makes

$$D_{4,2} = \begin{pmatrix} 1 & 4 & 0 \\ 0 & 3 & 0 \\ 0 & -6 & 1 \end{pmatrix}^{-1} = \begin{pmatrix} 1 & -4/3 & 0 \\ 0 & 1/3 & 0 \\ 0 & 2 & 1 \end{pmatrix} = M_{4,2}^{-1}.$$

To find the values of the basic variables, we multiply the various rows of $D_{4,2}$ with the first column of M, and obtain $x_4 = 2/3$, $x_2 = 1/3$, $z_{00} = 2$.

We must also know which z_{0j} are negative, if any, and therefore we multiply the last row of $D_{4,2}$ by the several columns of M. This gives

$$2, \quad -1, \quad 0, \quad 2, \quad 0, \quad 2, \quad 1.$$

The first value is z_{00}, which we have already obtained. The zeros refer to the basic variables x_2 and x_4. We see that z_{01} is negative, therefore x_1 must be made basic.

At this stage we need the x_1 column of the tableau, and we obtain it by multiplying the rows of $D_{4,2}$ by the column of x_1 in M. This gives

$$\frac{5}{3}$$
$$\frac{1}{3}$$
$$-1.$$

The ratios which determine the variable to be removed from the basic set are

$$\tfrac{2}{3} \div \tfrac{5}{3} = \tfrac{2}{5}$$

and

$$\tfrac{1}{3} \div \tfrac{1}{3} = 1.$$

Therefore x_4 will be exchanged for x_1. We compute $M_{1.2}^{1}$ from that portion of the tableau which we know, i.e.

$$
\begin{array}{cc}
 & \begin{array}{ccc} x_4 & x_5 \end{array} \\
\begin{array}{c} x_4 \\ x_2 \\ {} \end{array} &
\begin{pmatrix} 2/3 & 1 & -4/3 & 0 \\ 1/3 & 0 & 1/3 & 0 \\ 2 & 0 & 2 & 1 \end{pmatrix}
\end{array}
\quad \text{and} \quad
\begin{array}{c} x_1 \\ {} \\ {} \end{array}
\begin{pmatrix} 5/3^* \\ 1/3 \\ -1 \end{pmatrix},
$$

to obtain

$$
D_{1,2} = \begin{pmatrix} 3/5 & -4/5 & 0 \\ -1/5 & 3/5 & 0 \\ 3/5 & 6/5 & 1 \end{pmatrix},
$$

which is, of course, the inverse of

$$
\begin{pmatrix} 3 & 4 & 0 \\ 1 & 3 & 0 \\ -3 & -6 & 1 \end{pmatrix}.
$$

At this stage we obtain the values of the basic variables, and of the objective function, by multiplying the rows of $D_{1.2}$ with the first column of M. Thus

$$x_1 = \tfrac{2}{5}, \qquad x_2 = \tfrac{1}{5}, \qquad z_{00} = \tfrac{12}{5},$$

and the various z_{0j} are obtained by multiplying the last row of $D_{1.2}$ with the columns of M. We have

$$\tfrac{12}{5}, \quad 0, \quad 0, \quad 1, \quad \tfrac{3}{5}, \quad \tfrac{6}{5}, \quad 1.$$

All these are non-negative, and hence we have reached the final solution.

5–5 Constructive proof of the duality and existence theorems. Let us consider, once more, Examples 5–1 and 5–2. These two problems are dual. For the sake of a more convenient comparison of tableaux in dual problems we change the notation slightly. We use the subscripts of the main variables in one problem as those of the slack variables in the other. Examples 5–1 and 5–2 read then

$$3x_1 + 4x_2 + x_3 \leqslant 2,$$
$$x_1 + 3x_2 + 2x_3 \leqslant 1,$$
$$x_i \geqslant 0.$$
Maximize $3x_1 + 6x_2 + 2x_3 = B.$

$$3y_4 + y_5 \geqslant 3,$$
$$4y_4 + 3y_5 \geqslant 6,$$
$$y_4 + 2y_5 \geqslant 2,$$
$$y_j \geqslant 0.$$
Minimize $2y_4 + y_5 = C.$

We introduce, respectively, the slack variables x_4, x_5, and y_1, y_2, y_3. The final tableaux are, in our present notation,

	x_4	x_5	x_3			y_1	y_2	
x_1	0.6	−0.8	−1	0.4	y_4	−0.6	0.2	0.6
x_2	−0.2	0.6	1	0.2	y_5	0.8	−0.6	1.2
					y_3	1.0	−1.0	1.0
B	0.6	1.2	1	2.4	C	−0.4	−0.2	2.4

The relationship between these two tableaux will be clear from inspection. The subscripts of the basic variables in one are those of the slack variables in the other tableau. The values of the two objective functions are the same. The values of the variables in one are those in the bottom row of the other, but with changed sign for those of the objective function to be minimized; otherwise it could not be the final tableau. The remainder of the tableau contains also the same values in both, with changed signs and transposed (i.e. their position reflected in the diagonal).

We shall call two tableaux which exhibit these relative features *dual tableaux*, irrespective of whether they are final, and hence irrespective of whether all values of the variables as shown in the last columns are non-negative.

We shall show that if one of such a pair of tableaux represents a system of equations which has originated from one of a dual system of inequalities, then the other represents a system which can have originated from the dual system of inequalities. This will lead to a proof of the Duality and the Existence theorems (Chapter 3) which is constructive, in that it shows how to arrive at the respective optimal solutions, if they exist, and how to find out if they do not.

Let us take a Simplex tableau, not necessarily the final one. For the sake of argument, let the problem be that of minimizing

$$C = c_1 x_1 + \ldots + c_n x_n,$$

subject to

$$a_{1j}x_1 + \ldots + a_{nj}x_n - x_{n+j} = b_j \qquad (j = 1, \ldots, m)$$

and

$$x_i \geqslant 0 \qquad (i = 1, \ldots, n + m = N).$$

We assume that the order of subscripts is such that the following tableau is valid:

	x_{m-s+1} $\cdots$	x_n	x_{n+s+1} $\cdots$	x_{n+m}	
x_1	$z_{1,m-s+1}$	z_{1n}	$z_{1,n+s+1}$	$z_{1,n+m}$	x_1^0
$\cdot$					
$\cdot$					
$\cdot$					
x_{m-s}	$z_{m-s,m-s+1}$	$z_{m-s,n}$	$z_{m-s,n+s+1}$	$z_{m-s,n+m}$	x_{m-s}^0
x_{n+1}	$z_{n+1,m-s+1}$	$z_{n+1,n}$	$z_{n+1,n+s+1}$	$z_{n+1,n+m}$	x_{n+1}^0
$\cdot$					
$\cdot$					
$\cdot$					
x_{n+s}	$z_{n+s,m-s+1}$	$z_{n+s,n}$	$z_{n+s,n+s+1}$	$z_{n+s,n+m}$	x_{n+s}^0
C	$z_{0,m-s+1}$	$z_{0,n}$	$z_{0,n+s+1}$	$z_{0,n+m}$	z_{00}

We repeat for the reader's convenience the following identities, as applied to the present case:

$$x_i^0 = \sum_{t=1}^{m} d_{ti}b_t \tag{2-a}$$

$$z_{ik} = \sum_{t=1}^{m} d_{ti}a_{kt} \qquad \text{for } k = m - s + 1, \ldots, n \tag{2-b}$$

$$= -d_{k-n,i} \qquad \text{for } k = n + s + 1, \ldots, n + m$$
$$\text{(because any } x_{n+j} \text{ appears only in the } j\text{th equation)}$$

$$z_{00} = \sum_{i=1}^{m-s} c_i x_i^0 \tag{2-c}$$

$$z_{0t} = \sum_{i=1}^{m-s} c_i z_{it} - c_t \qquad \text{for } t = m - s + 1, \ldots, n \tag{2-d}$$

$$= \sum_{i=1}^{m-s} c_i z_{it} \qquad \text{for } t = n + s + 1, \ldots, n + m.$$

All x_i^0 $(i = 1, \ldots, m - s, n + 1, \ldots, n + s)$ are non-negative. If this is the final tableau, then all z_{0t} $(t = m - s + 1, \ldots, n, n + s + 1, \ldots, n + m)$ are non-positive.

Introduce now the tableau which is dual to the one just considered:

	y_1 $\cdots$	y_{m-s}	y_{n+1} $\cdots$	y_{n+s}	
y_{m-s+1}	$-z_{1,m-s+1}$	$-z_{m-s,m-s+1}$	$-z_{n+1,m-s+1}$	$-z_{n+s,m-s+1}$	$-z_{0m-s+1}$
.					
.					
.					
y_n	$-z_{1n}$	$-z_{m-s,n}$	$-z_{n+1,n}$	$-z_{n+s,n}$	$-z_{0,n}$
y_{n+s+1}	$-z_{1,n+s+1}$	$-z_{m-s,n+s+1}$	$-z_{n+1,n+s+1}$	$-z_{n+s,n+s+1}$	$-z_{0,n+s+1}$
.					
.					
y_{n+m}	$-z_{1,n+m}$	$-z_{m-s,n+m}$	$-z_{n+1,n+m}$	$-z_{n+s,n+m}$	$-z_{0n+m}$
B	x_1^0	x_{m-s}^0	x_{n+1}^0	x_{n+s}^0	z_{00}

We want to find out whether this tableau could arise during a solution of the problem

$$\text{Maximize } b_1 y_{n+1} + \ldots + b_m y_{n+m} = B, \text{ say,}$$

subject to

$$a_{i1} y_{n+1} + \ldots + a_{im} y_{n+m} + y_i = c_i \qquad (i = 1, 2, \ldots, n),$$
$$y_j \geqslant 0 \qquad\qquad\qquad\qquad\quad (j = 1, \ldots, N).$$

In view of the condition that the y_j must be non-negative, this is only possible within the Simplex routine if

$$-z_{0,m-s+1}, \ldots, -z_{0,n}, \qquad -z_{0,n+s+1}, \ldots, -z_{0,n+m}$$

are all non-negative. This will be so, if the tableau in the primal problem is the final one, because then all the $z_{0,m-s+1}$ and so on will be non-positive. At any previous stage this will not be so. However, let us, for the time being, ignore the condition that we want to deal only with feasible solutions, and let us go on to show that the tableau represents relations which are transformations of the given conditions of the dual problem, and of the definition of B.

In other words, we show that the constraints are identically satisfied by

$$\text{(i)} \quad y_j = \sum_{t=1}^{m-s} z_{tj} y_t + \sum_{t=n+1}^{n+s} z_{tj} y_t - z_{0j}$$

for all basic y-variables and also that

$$\text{(ii)} \quad b_{s+1} z_{0,n+s+1} + \ldots + b_m z_{0,n+m} = -z_{00}.$$

It is convenient to write the nonbasic variables also in the form (i); this is possible by defining

$$z_{0j} = 0 \qquad \text{for } j = 1, \ldots, m - s, n + 1, \ldots, n + s,$$

$$z_{tj} = \begin{cases} 0 & \text{when } t \neq j, \\ 1 & \text{when } t = j. \end{cases}$$

The relation $z_{0t} = \sum_{i=1}^{m-s} c_i z_{it} - c_t$ remains valid.

(i) We must show that

$$\sum_{j=n+1}^{n+m} a_{i,j-n} \left[\sum_{t=1}^{m-s} z_{tj} y_t + \sum_{t=n+1}^{n+s} z_{tj} y_t - z_{0j} \right]$$
$$+ \sum_{t=1}^{m-s} z_{ti} y_t + \sum_{t=n+1}^{n+s} z_{ti} y_t - z_{0i} = c_i$$

for $i = 1, \ldots, n$.

The algebraic proof is quite straightforward. Using the definition of z_{0t} we write, for the left-hand side,

$$\sum_{j=n+1}^{n+m} a_{i,j-n} \left[\sum_{t=1}^{m-s} z_{tj} y_t + \sum_{t=n+1}^{n+s} z_{tj} y_t \right] - \sum_{j=n+1}^{n+m} a_{i,j-n} \sum_{t=1}^{m-s} c_t z_{tj}$$
$$+ \sum_{t=1}^{m-s} z_{ti} y_t + \sum_{t=n+1}^{n+s} z_{ti} y_t - \sum_{t=1}^{m-s} c_t z_{ti} + c_i,$$

or

$$\left(\sum_{t=1}^{m-s} y_t + \sum_{t=n+1}^{n+s} y_t - \sum_{t=1}^{m-s} c_t \right) \left(\sum_{j=n+1}^{n+m} a_{i,j-n} z_{tj} + z_{ti} \right) + c_i.$$

But by definition

$$\sum_{j=n+1}^{n+m} a_{i,j-n} z_{tj} = - \sum_{j=n+1}^{n+m} a_{i,j-n} d_{j-n,t} = -z_{ti}.$$

Hence the expression reduces to c_i.

(ii) $$\sum_{=n+1}^{n+m} b_{j-n} z_{0j} = \sum_{j=n+1}^{n+m} b_{j-n} \sum_{t=1}^{m-s} c_t z_{tj}$$
$$= \sum_{t=1}^{m-s} c_t \sum_{j=n+1}^{n+m} b_{j-n}(-d_{j-n,t}) = - \sum_{t=1}^{m-s} c_t x_t^0 = -z_{00}.$$

We have thus proved that the second tableau can arise as representing a solution to the dual problem, though not necessarily a feasible one. If the first tableau is final, then the values of the y variables are all non-negative, and so are then the values x_i^0 in the bottom row of the dual tableau; the latter is, hence, also final. The same value z_{00}

appears in both final tableaux, and we have thus proved the Existence theorem as well as the Duality theorem.

We have, in fact, proved more, namely the complete relationship between final tableaux of two dual problems.

We have seen in Section 3–5 that the spaces defined by two dual systems, of n and m dimensions respectively, are orthogonal in a space of $m + n$ dimensions, in which they are both embedded. If we have two dual tableaux, we can write the two problems in such a way that the basic variables are slack, and the nonbasic variables define the two orthogonal spaces. The Simplex, or equivalently the Dual Simplex, method (see Section 5–6) are procedures which redefine the two spaces until eventually all slack variables have non-negative values.

5–6 Dual simplex method. We have seen that to each tableau of the Simplex method there corresponds a dual tableau, and that to an optimal tableau of a problem there corresponds an optimal tableau of the dual problem.

To fix our ideas, assume that the primal problem is one of minimizing. In that case in the Simplex tableau all variables will have non-negative values, and the values of the z_{0j} might have any sign, until in the final tableau they will all be non-positive. In the succession of the dual tableaux, referring to the dual problem, the z_{0j} will all be non-negative, but the values of the basic variables will only be all non-negative in the final tableau.

This suggests a procedure, dual and alternative to the Simplex method, which we call the *Dual Simplex method*. It is due to C. E. Lemke [99].* We start with some *dual feasible* tableau, where all z_{0j} have the right sign, but without insisting on feasibility of the variables (though still restricting ourselves to basic solutions). Whenever we have a negative basic variable, we transform the tableau. The rules of proceeding are simply a translation of those for the Simplex method, and read as follows:

Choose any variable x_r with negative value to be made nonbasic.

Consider all *negative* values z_{rj} in that row to be eligible as pivot, and compute $|z_{0j}/z_{rj}|$. The smallest of these ratios decides the nonbasic variable x_h, to be made basic.

The transformation rules are the same as in the Simplex method, since they express the process of elimination and substitution, which is independent of the sign of the variables involved.

We have introduced the Dual Simplex method as a procedure consisting of steps derived from the Simplex method. This has suggested

---

* Some writers prefer to call it the Dual method.

a parallelism of tableaux in the two methods, such that the objective functions had the same value at each step. If the problem to be solved for the primal is one of maximizing, and that of the dual one of minimizing, then the values of the objective functions will increase, simultaneously, in both problems. This might look surprising, considering that in the second problem we want to minimize. It is explained by the fact that, up to the final tableaux, we are dealing in the dual problem with solutions which are not feasible. We have overshot the target, and must trace our steps back to a feasible solution, which is then optimal.

The remarks which now follow refer to difficulties which we had to deal with in connection with the Simplex method as well, such as finding a first solution, degeneracy, and so on. Knowledge of criteria and devices in the Simplex method will help us here. As a general principle, we shall have to remember that variables in one problem correspond to constraints in its dual.

Thus the task of finding a first solution is easily tackled. This time, it does not have to be feasible; we are looking, instead, for a basic solution such that, if we express the objective function in terms of the nonbasic variables, they have already the signs required.

We remember the device of the Simplex method which used just one artificial variable, and choose the dual procedure. That is, we introduce an artificial constraint

$$x_1 + \ldots + x_n \leqslant M,$$

or

$$x_1 + \ldots + x_n + x_0 = M,$$

where x_0 must also be non-negative.

EXAMPLE 5-8. Take Example 5-1. We solve it now by the Dual Simplex method (note the changed subscripts).

$$3x_3 + 4x_4 + x_5 + x_1 = 2,$$
$$x_3 + 3x_4 + 2x_5 + x_2 = 1.$$
$$\text{Maximize } 3x_3 + 6x_4 + 2x_5 = B, \text{ say.}$$

Here all the coefficients in the objective function have the wrong sign. Therefore we add the constraint

$$x_3 + x_4 + x_5 + x_0 = M.$$

The largest coefficient with wrong sign in B is x_4, so we write

$$x_4 = M - x_3 - x_5 - x_0,$$

and substitute into B. Thus

$$B = 6M - 3x_3 - 4x_5 - 6x_0.$$

Now B is suitable as a starting point for the Dual Simplex method. We substitute, also, into the constraints

$$- x_3 - 3x_5 + x_1 - 4x_0 = 2 - 4M,$$
$$-2x_3 - x_5 + x_2 - 3x_0 = 1 - 3M.$$

We can take x_1, x_2, and x_4 as the basic variables and have the tableau

	x_0	x_3	x_5	
x_1	-4	-1	-3	$2 - 4M$
x_2	-3	-2	-1	$1 - 3M$
x_4	1	1	1	M
B	6	3	4	$6M$

which is precisely the dual of the first tableau for its dual (Ex. 5–2), when solved by the Simplex method. We need not continue with it here. The variable x_0 will be dropped when it becomes basic.

We consider now criteria for inconsistency, or for unbounded solutions, in the Dual Simplex method. Unboundedness in the primal is equivalent to inconsistency in the dual. (See Section 3–4.) In the Dual Simplex method we would thus find inconsistency indicated if we cannot find any negative coefficient eligible to become a pivot, in the row of any variable with negative value. This is also understandable from first principles: if, say,

$$x_1 + \sum_i a_{i1} x_i = x_1^0 < 0 \qquad \text{and all} \qquad a_{i1} \geqslant 0,$$

then no non-negative values of the variables can possibly satisfy the inequality.

EXAMPLE 5–9. (Compare with Example 5–2.) Given

$$3x_1 + x_2 - x_3 = 3,$$
$$4x_1 + 3x_2 - x_4 = 6,$$
$$x_1 + 2x_2 - x_5 = 2.$$
$$\text{Maximize } 2x_1 + x_2.$$

The additional constraint called for is

$$x_1 + x_2 + x_0 = M,$$

which leads to

$$x_1 = M - x_2 - x_0,$$

so that we have the tableau

	x_0	x_2	
x_3	3	2	$3M - 3$
x_4	4	1	$4M - 6$
x_5	1	-1	$M - 2$
x_1	1	1	M
B	2	1	$2M$

As it turns out, this is already the final tableau. M is still contained in it, so that the variables have unbounded values.

5-7 Dual simplex method. Cycling. In the Simplex method the possibility of cycling arose from too many hyperplanes passing through the same point. The analogous situation in the Dual Simplex method arises when there are lines connecting intersections of $N - m$ hyperplanes which are perpendicular to the preferred direction. Moving along such a line does not improve the objective function.

EXAMPLE 5–10. Given
$$y_3 + 2y_4 + y_5 \geqslant 1,$$
$$-2y_3 - 3y_4 + 4y_5 \geqslant 1.$$
$$\text{Minimize } y_5.$$

To interpret this problem geometrically, we introduce slack variables y_1 and y_2 and obtain the diagram of Fig. 5–3. The planes $y_1 = 0$ and $y_2 = 0$ intersect along the line PQ, where $P = (0, 3/11, 5/11)$ and

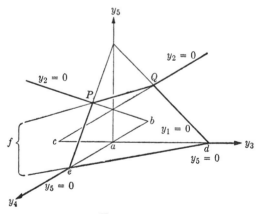

FIGURE 5–3

$Q = (1/2, 0, 1/2)$. Of these P has the smaller y_5 coordinate, and is thus the point whose coordinates are optimal. (The feasible region is behind the plane $y_2 = 0$, as seen in this graph.)

The present problem is dual to that of Example 5–6. The letters a to f attached to the points in the graph above are the same as those attached to the corresponding descriptions of one and the same point in the earlier example. In the latter we obtained the answer, by the Simplex method, by moving through a, d, f, and eventually to a point corresponding to P.

We want to solve the present problem by the Dual Simplex method. The successive tableaux will be dual to those in the previous case, and we shall move, again, through the points a, d, f, and P. The first three are now not feasible. It is of interest to see how their succession can be explained.

We start with a. This is not a feasible point, and hence not the final answer. However, not every nonfeasible point could appear in the Dual Simplex method. It must have the property that (assuming we want to minimize, for the sake of argument) the expression of the objective function in terms of the nonbasic variables has positive coefficients (negative values in the tableau). Geometrically, it must be a point which is farthest in the preferred direction in that polyhedral cone (in two dimensions: in that angle) which would be the feasible region if all basic variables and their hyperplanes were ignored. The point a has this property in the cone defined by $y_3 = y_4 = y_5 = 0$.

Turning now to the remaining variables, we see that a is not feasible because its coordinates y_1 and y_2 are negative. We must therefore move along an edge to a point where one of these features is remedied. We could move either to d or to e. In the Simplex method this decision depends on a perturbation. Here perturbation means a slight tilting of the preferred direction, say from the negative y_5 axis forward and to the right. We move within the cone in a direction with a negative component in the preferred direction (because we have overshot and must increase the objective function) and this decides in favour of d. It is again a lowest point in its cone, i.e. in that defined by $y_1 = y_4 = y_5 = 0$. It is not feasible, because its y_2 coordinate is negative, and therefore we move towards the plane $y_2 = 0$. This brings us to f, which is eligible, though not feasible. Finally, moving to P brings us out of the plane $y_5 = 0$, and to the optimal point. If we had chosen a different perturbation, our progress might have been from a through e directly to P.

It will be noticed, incidentally, that we do not always move to an adjacent point. Although e is adjacent to d, it was passed over in favour of f.

There is no difficulty in finding an example for cycling. We simply take the dual of that example which showed cycling in the Simplex algorithm, and solve it by the Dual Simplex method. Throughout the cycling, we remain on the same hyperplane.

The study of cycling in the Dual Simplex algorithm has the advantage that in the geometrical representation we move from point to point and can thus more easily survey the progress than in the Simplex method, where we remain in the same point, though this point is described in different ways.

If we consider the case where there are two zero coefficients in the expression of the objective function in terms of the nonbasic variables, then that part of the procedure which leaves the value of the objective function unchanged can be represented in a plane. For cycling to take place we would have to remain all the time in this plane.

Now imagine that we are at a point α (Fig. 5–4) and move from there to β. We do this because there is another edge through β, and we were

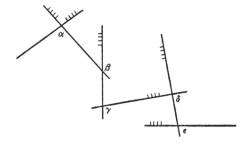

FIGURE 5–4

on its wrong side. From β, we must move along this other edge, until we are stopped again, in γ. The second edge there must take us away from the side on which α lay, so that α and δ must be on different sides of the edge through β and γ. Similarly, proceeding to ε, β and ε must lie on different sides of the edge through γ and δ.

The same must be true of any four successive points. If we imagine that we have reached a certain point P, and then turn to the right, say, then at the next point we turn left, and so on alternately. Hence a cycle must consist of an even number of points, because otherwise, when we return to P, we would this time turn left, which contradicts the geometrical rules governing our progress.

It follows that α and ε cannot be the same point. But a further point ζ can serve to close the cycle. This is, in fact, what happens in the example of cycling given earlier in this chapter. Our argument shows that there must be at least six steps in a cycle (Fig. 5–5).

We repeat that the argument applies to the case with two zeros in the expression of the objective function. Our presentation is based on [25], where it is also shown that with only one zero in that expression cycling is impossible. This follows also easily from our consideration of perturbations in an earlier section.

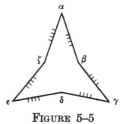

FIGURE 5–5

5–8 Bounded variables. In some cases it is possible to take advantage of the special structure of some at least of the constraints to simplify the tableau. This idea has been applied, in particular, to the case of *bounded variables*, i.e. when there exist constraints $x_i \leqslant u_i$, where the u_i are given constants. (If there are also lower bounds, say, v_i, then we introduce new variables $x'_i = x_i - v_i$, and have again a problem with non-negative variables.) Even if these upper bounds appear only for some variables, we can always introduce sufficiently high bounds for all of them without thereby changing the problem, provided the solutions of the constraints are bounded.

A simplification of the Simplex method has been worked out by G. B. Dantzig in [41] and by A. Charnes (in a Res. Mem. of the Carnegie Institute of Technology). We discuss here a method due to H. M. Wagner (see [122]).

Let the given system be

$$\sum_{i=1}^{n} a_{ij}x_i + x_{n+j} = b_j \qquad (j = 1, \ldots, m),$$

$$x_i + x'_i = u_i \qquad (i = 1, \ldots, n),$$

$$x_i, x'_i, x_{n+j} \geqslant 0.$$

$$\text{Maximize } c_1x_1 + \ldots + c_nx_n.$$

We may assume that all c_i are negative or zero. For those i where they are not, we substitute x'_i for x_i and obtain an equivalent system where all coefficients of the variables in the objective function are non-positive. Only the corresponding variables appear in the tableau. We obtain a starting solution by making the slack variables basic. If they are all non-negative, then the optimum has been reached. Otherwise we carry on with the steps of the Dual Simplex method, remembering the existence of upper bounds. The choice of the pair of variables to be exchanged remains unaffected. If after transformation we find that a variable exceeds its upper limit, then we replace it by its complement, with the corresponding changes of signs in the tableau. The complementary variable is then negative, and we apply again the Dual Simplex algorithm. When we reach the optimal tableau for its $m + n$ variables, we have also reached the optimum for all $2n + m$ variables. A reader who feels that this should be rigorously shown will be able to supply the proof himself. (As an example, see the first part of Exercise 10–4.)

A method closely related to the Dual Simplex method has been introduced by E. M. L. Beale in [23]. He calls it the *Method of Leading Variables*.

There exist methods, usually referred to as *Composite Algorithms*, which make restricted use of the dual method to overcome an initial difficulty of finding a feasible solution. In one version one uses the slack variables to determine a first, not necessarily feasible, solution, by making these variables equal to the right-hand side, and introduces a fictitious objective function with coefficients which have the correct sign, so that the Dual Simplex method can be applied. Having found the optimum, we have a feasible solution of the constraints, and can then revert to that objective function which we are interested in.

A device, related to this, can be used when a problem has been solved and another problem arises which differs from the earlier one only in the value of one of the constants on the right-hand side of a constraint, perhaps because the first problem contained this constant in error. It has been pointed out by H. Markowitz, in a report of the RAND corporation, that if the slack variable of that constraint is basic, then it will simply undergo the same change as that constant, and if it becomes thereby negative, then an application of the Dual Simplex method is called for. If the slack variable in question is non-basic, however, then it will be increased, or decreased, until some other variable is zero. In any case, an exchange of the two variables is indicated, followed in the latter by an application of the Dual Simplex method. The reader is warned not to confuse such methods with the Primal-dual algorithm, explained later in this chapter.

5–9 The multiplex method. In [69] R. Frisch has developed an algorithm which is not restricted to the consideration of basic solutions only and which might, in conjunction with other devices mentioned in his paper, lead sometimes more quickly to the final answer than other methods.

Let equality constraints in N variables be given. Start by finding some feasible, not necessarily basic, solution and express all N variables in terms of $N - m = n$ of them. The latter are called *basis-variables*. If their values are all zero, then they correspond to what we called nonbasic variables. However, the set of basis-variables remains the same throughout the computations.

Proceed in the preferred direction until a boundary has been reached. This boundary may happen to be of any dimension from 0 (a vertex) to $n - 1$ (a hyperplane). The next step consists in using a regression technique which—in geometric terms—projects the preferred direction on to the boundary space. If that projection has a positive component within the feasible region, proceed in that direction. Otherwise it becomes necessary to enlarge the boundary space by judiciously ignoring some of its defining constraints. The paper contains a detailed description of the relevant criteria, and of those indicating that the optimum has been reached.

The ideas of this method may be applied to nonlinear cases (objective functions and/or constraints) as well, and are comparable to methods proposed by J. Rosen (see Chapter 12).

5–10 The cross section method. Yet another method may be mentioned here. It is due to Jeremy J. Stone, who calls it the *Cross Section method*.

We start with the following simple remark: If we have only one single constraint, say

$$a_1 x_1 + \ldots + a_n x_n = a \qquad (> 0),$$

apart from $x_i \geqslant 0$ for all i, with the objective function

$$c_1 x_1 + \ldots + c_n x_n = C, \text{ say,}$$

to be minimized, and if we know that the feasible region is unbounded, then we have only one single basic variable. If it is x_k, then $x_k = a/a_k$ and $C = c_k \times a/a_k$. Therefore we choose that k for which c_k/a_k is smallest, out of all those k for which $a_k > 0$.

EXAMPLE 5–11. (Same as Example 5–2.) Given

$$3x_1 + x_2 - x_3 = 3,$$
$$4x_1 + 3x_2 - x_4 = 6,$$
$$x_1 + 2x_2 - x_5 = 2,$$

Minimize $2x_1 + x_2 = C$, say.

The Cross Section method consists of taking, to begin with, just one constraint, e.g. the first, and minimizing C. This is done by

$$x_1 = 1 - \frac{x_2}{3} + \frac{x_3}{3}, \qquad x_2 = 0, \qquad x_3 = 0, \qquad C = 2 + \frac{x_2}{3} + \frac{2x_3}{3}.$$

Now we introduce the second constraint as well. The idea of this method consists of retaining x_1 in the basis, and finding another variable which, together with x_1, is a feasible basis for the enlarged problem. Such a variable must exist; this can be seen as follows: Starting with one constraint and then introducing another is equivalent, in the dual problem, to starting with one variable, and then introducing another. Now if a feasible solution to the problem with just one variable has been found, then a feasible solution to the two-variable case is immediately given by retaining the value of the first, and making the second zero. However, even if the first solution was optimal, there is no reason why the second should be. We must therefore, usually, apply iterative steps to obtain the optimum for this phase. Similar remarks apply to the progress from k to $k + 1$ variables, for any k.

In geometric terms, consider that we have already reached an optimal solution for a two-variable case. For the dual, this means that a vertex has been found of a polygon in two dimensions. If we add a further variable, then we have a third dimension to consider. The intersection of the two straight lines determining that vertex remains feasible, but to obtain the new, optimal vertex, we might have to slide along the straight-line intersection (or cross section, hence the name of the method) of the new planes along which the previous vertex lies.

Stone has constructed an algorithm for finding the new basis. The merits claimed for this method are, to begin with, that it is easy to find a first solution, and hence one starts at a point which would, perhaps, only be reached later in the Simplex method. It is possible, though, that a problem with fewer constraints has no finite solution, while the complete problem has. In that case one introduces an artificial constraint, in the usual way.

A further advantage of the method is the possibility of discovering various features of the problem as one goes along, for instance, redundancy of certain constraints for the determination of the optimum.

5–11 The primal-dual algorithm. We introduce now a computational algorithm which is particularly useful in connection with special cases, as will be seen in Chapter 6. It is due to G. B. Dantzig, L. R. Ford, and D. R. Fulkerson [49].

The fundamental principle of this algorithm can be described as follows. We start from a basic feasible solution of a system which we

a

describe as the dual system. To this solution there corresponds a basic solution of the primal, which we write as a system of equations, using slack variables. We can find this corresponding solution by setting those variables equal to zero which correspond to those constraints in the dual problem which are satisfied as strict inequalities. This solution to the primal will only be feasible, though, when we have reached the final stage. So we construct an *extended primal*, introducing artificial variables as in the Two-phase method, and minimize the sum of the artificial variables. The procedure will best be explained by an example.

EXAMPLE 5–12. (Examples 5–1 and 5–2 combined.) Given

$$3x_1 + x_2 - x_3 = 3, \qquad 3y_3 + 4y_4 + y_5 \leqslant 2,$$
$$4x_1 + 3x_2 - x_4 = 6, \qquad y_3 + 3y_4 + 2y_5 \leqslant 1,$$
$$x_1 + 2x_2 - x_5 = 2, \qquad -y_3 \leqslant 0,$$
$$x_i \geqslant 0. \qquad\qquad -y_4 \leqslant 0,$$
$$-y_5 \leqslant 0.$$

Minimize $C = 2x_1 + x_2$.   Maximize $B = 3y_3 + 6y_4 + 2y_5$.

We have written the constraints of the primal system as equations, so that the variables in the dual are not sign-restricted; but the form of the inequalities indicates that in fact the sign restriction holds.

We start off with some solution of the dual. If none were at once apparent, we could introduce an artificial constraint with a new variable, as has been explained. Let the solution be $y_3^0 = y_4^0 = y_5^0 = 0$. Then the first two constraints of the dual system are satisfied as inequalities. We therefore make x_1 and x_2 zero in the primal. However, then the equations of the primal would not have a feasible solution. So we introduce artificial variables, and write the extended primal

$$3x_1 + x_2 - x_3 + x_{101} = 3,$$
$$4x_1 + 3x_2 - x_4 + x_{102} = 6,$$
$$x_1 + 2x_2 - x_5 + x_{103} = 2,$$
$$x_i \geqslant 0 \quad \text{for all } i,$$

and we aim at minimizing $x_{101} + x_{102} + x_{103}$. When this sum equals zero, then all the artificial variables are also zero, and the original constraints are restored.

We put x_1 and x_2 equal to zero in the extended primal, and obtain the *restricted primal*

$$-x_3 + x_{101} = 3,$$
$$-x_4 + x_{102} = 6,$$
$$-x_5 + x_{103} = 2,$$
$$x_i \geqslant 0.$$

Minimize $C' = x_{101} + x_{102} + x_{103}$.

The solution is, of course, $x_{101} = 3$, $x_{102} = 6$, $x_{103} = 2$, $C' = 11$. Consider, now, the dual to the restricted primal

$$0 \leqslant z_1 \leqslant 1, \qquad 0 \leqslant z_2 \leqslant 1, \qquad 0 \leqslant z_3 \leqslant 1.$$

Maximize $3z_1 + 6z_2 + 2z_3$.

The solution is found by considering those constraints to be equations, which correspond to the basic variables in the restricted primal. Hence $z_1^0 = z_2^0 = z_3^0 = 1$.

This terminates one step. We look now for an improved solution of the dual. We assert that:

(i) $y_i^0 + t z_{i-2}^0$ $(i = 3, 4, 5)$ is also a feasible solution of the original dual for any positive t less than a t_0 to be ascertained.

(ii) The objective function of the dual increases with increasing t.

It will now be proved that these assertions are generally true. (i) Let the extended primal be

$$a_{1j}x_1 + \ldots + a_{Nj}x_N + x_{M+j} = b_j \geqslant 0 \qquad (j = 1, \ldots, m),$$
$$C' = \sum_j x_{M+j},$$
$$\text{all } x_i \geqslant 0.$$

The original dual is

$$a_{i1}y_{N+1} + \ldots + a_{i,m}y_{N+m} + y_i = c_i \qquad (i = 1, \ldots, N),$$
$$B = b_1 y_{N+1} + \ldots + b_m y_{N+m},$$

with $y_1, \ldots, y_N$ non-negative, while y_{N+j} may have any sign for $j = 1, \ldots, m$.

Let this system have a solution with

$$y_1^0 = \ldots = y_k^0 = 0, \qquad y_{k+1}^0 > 0, \ldots, y_N^0 > 0.$$

The restricted primal is then

$$a_{1j}x_1 + \ldots + a_{kj}x_k + x_{M+j} = b_j \qquad (j = 1, \ldots, m),$$
$$C' = x_{M+1} + \ldots + x_{M+m}.$$

The dual to this system is

$$a_{11}z_1 + \ldots + a_{1m}z_m \leqslant 0,$$

$$\cdot$$
$$\cdot$$
$$\cdot$$

$$a_{k1}z_1 + \ldots + a_{km}z_m \leqslant 0,$$

$$z_1 \leqslant 1,$$

$$\cdot$$
$$\cdot$$
$$\cdot$$

$$z_m \leqslant 1,$$

$$B' = b_1z_1 + \ldots + b_mz_m.$$

Let it have a solution $z_1^0, \ldots, z_m^0$. Our statement (i) means that if

$$a_{i1}y_{N+1}^0 + \ldots + a_{im}y_{N+m}^0 \leqslant c_i \quad \text{for} \quad i = 1, \ldots, N,$$

then there exists a positive t such that also

$$a_{i1}(y_{N+1}^0 + tz_1^0) + \ldots + a_{im}(y_{N+m}^0 + tz_m^0) \leqslant c_i \quad \text{for the same } i.$$

The inequalities remain valid for $i = 1, \ldots, k$ because the left-hand side has been increased by $t(a_{i1}z_1^0 + \ldots + a_{im}z_m^0)$, and this is smaller than or equal to 0 for $i = 1, \ldots, k$. For $i = k + 1, \ldots, N$, on the other hand, the inequality was strictly valid for

$$a_{i1}y_{N+1}^0 + \ldots + a_{im}y_{N+m}^0,$$

and hence remains valid as long as

$$t \leqslant \frac{y_i^0}{a_{i1}z_1^0 + \ldots + a_{im}z_m^0} \quad \text{for} \quad i = k + 1, \ldots, N,$$

i.e. as long as t does not exceed the smallest of the right-hand side values; we call this minimum t_0. Of course there may not be any $i > k$. Then the solutions for the dual are unbounded, and hence the primal has no solution.

(ii) The value of the objective function of the original dual problem changes with the new solution from $\sum_j b_j y_{N+j}^0$ to

$$\sum_j b_j y_{N+j}^0 + t\sum_j b_j z_j^0 = \sum_j b_j y_{N+j}^0 + t\sum_j x_{M+j}$$

(by the principle of duality), and as long as the last sum is positive, this means a strict increase. On the other hand, when

$$\sum_j x_{M+j} = 0,$$

then we have reached the optimum.

EXAMPLE 5–12 (continued). We find t_0 as the largest of the t satisfying

$$8t \leqslant 2, \qquad 6t \leqslant 1,$$

i.e. $t_0 = 1/6$. The new solution to the dual is thus

$$y_3^0 = \tfrac{1}{6}, \qquad y_4^0 = \tfrac{1}{6}, \qquad y_5^0 = \tfrac{1}{6},$$

and the restricted primal is then, with its dual,

$$x_2 + x_{101} = 3,$$
$$3x_2 + x_{102} = 6,$$
$$2x_2 + x_{103} = 2, \qquad z_1 + 3z_2 + 2z_3 \leqslant 0,$$
$$x_i \geqslant 0. \qquad\qquad z_1 \leqslant 1, \quad z_2 \leqslant 1, \quad z_3 \leqslant 1.$$

Minimize $C' = x_{101} + x_{102} + x_{103}$ \qquad Maximize $B' = 3z_1 + 6z_2 + 2z_3.$

The solutions are, respectively,

$$x_{101} = 2, \qquad x_{102} = 3, \qquad x_{103} = 0; \qquad z_1 = 1, \qquad z_2 = 1, \qquad z_3 = -2;$$
$$x_2 = 1, \qquad C' = 5. \qquad\qquad B' = 5.$$

We find t_0 as the largest value satisfying

$$\tfrac{4}{3} + 5t \leqslant 2, \qquad -\tfrac{1}{6} + t \leqslant 0, \qquad -\tfrac{1}{6} + 2t \leqslant 0.$$

Hence
$$t_0 = \tfrac{1}{12}, \qquad y_3^0 = \tfrac{1}{4}, \qquad y_4^0 = \tfrac{1}{4}, \qquad y_5^0 = 0.$$

We continue, as follows:

$$x_2 + x_{101} = 3, \qquad\qquad z_1 + 3z_2 + 2z_3 \leqslant 0,$$
$$3x_2 + x_{102} = 6, \qquad\qquad -z_3 \leqslant 0,$$
$$2x_2 - x_5 + x_{103} = 2, \qquad\qquad z_1 \leqslant 1,$$
$$C' = x_{101} + x_{102} + x_{103}. \qquad\qquad z_2 \leqslant 1,$$
$$z_3 \leqslant 1,$$
$$B' = 3z_1 + 6z_2 + 2z_3.$$

The solution here is

$$x_{101} = 1, \quad x_{102} = 0, \qquad\qquad z_1 = 1, \quad z_2 = -\tfrac{1}{3}, \quad z_3 = 0,$$
$$x_{103} = 0, \quad x_2 = 2, \quad x_5 = 2,$$
$$C' = 1. \qquad\qquad\qquad\qquad B' = 1.$$

We find

$$t_0 = \tfrac{3}{20}, \quad y_3^0 = \tfrac{2}{5}, \qquad y_4^0 = \tfrac{1}{5}, \qquad y_5^0 = 0.$$

Restricted primal	Dual to restricted primal

$$\begin{array}{ll}
3x_1 + x_2 + x_{101} = 3, & 3z_1 + 4z_2 + z_3 \leqslant 0, \\
4x_1 + 3x_2 + x_{102} = 6, & z_1 + 3z_2 + 2z_3 \leqslant 0, \\
x_1 + 2x_2 - x_5 + x_{103} = 2, & -z_3 \leqslant 0, \\
& z_1 \leqslant 1, \quad z_2 \leqslant 1, \quad z_3 \leqslant 1, \\
C' = x_{101} + x_{102} + x_{103}. & B' = 3z_1 + 6z_2 + 2z_3.
\end{array}$$

The final solution is

$$x_1 = \tfrac{3}{5}, \qquad x_2 = \tfrac{6}{5}, \qquad x_5 = 1, \qquad C' = 0.$$

The objective function is $C = B = 12/5$.

5–12 Relaxation method. It is a requirement of all linear programming algorithms that at some stage a feasible solution must be found.
This problem, that is finding a set of values of the variables which
satisfy a number of inequalities, has been approached in various ways.
As an example, we mention here the Relaxation method [106]. Its
essential idea can be explained in terms of geometry, as follows.

Start with some arbitrary point. If its coordinates are not a solution,
then compute its distances from all those hyperplanes on whose wrong
sides the point lies. Take the hyperplane which is farthest away and
project the point on to it—in another version reflect the point in it—
and repeat the procedure. The paper mentioned studies the conditions
in which this procedure terminates in a finite number of steps with a
feasible point, converges to such a point, or oscillates.

EXERCISES

5–1. Solve the following problems by the Simplex method.

$$\begin{array}{ll}
x_1 + x_2 \leqslant 3, & \\
x_1 - 2x_2 \leqslant 1, & \text{All } x_i \text{ non-negative.} \\
-2x_1 + x_2 \leqslant 2. &
\end{array}$$

(a) Minimize $x_1 - x_2$. (b) Maximize $x_1 - x_2$.

5–2. Show that by omitting one constraint in Exercise 5–1 both (a) and (b) have an infinite solution.

5–3. Indicate those limits of t in the objective function $tx_1 - x_2$ beyond which either a finite minimum or a finite maximum exists when those constraints hold which were considered to remain in Exercise 5–2.

5–4. Solve the following problems.

$$\begin{aligned} -x_1 + 2x_2 - x_3 &= 1, \\ -x_1 - x_2 + 2x_3 &= 1. \end{aligned} \quad \text{All } x_i \text{ non-negative.}$$

(a) Maximize $2x_1 - x_2 - x_3$.   (b) Minimize $2x_1 - x_2 - x_3$.

5–5. Solve the following problem by the Simplex method.

$$\begin{aligned} -0.5x + 1.3y &\leqslant 0.8, \\ 4x + y &\leqslant 10.7, \\ 6x + y &\leqslant 15.4, \qquad x, y \geqslant 0. \\ 6x - y &\leqslant 13.4, \\ 4x - y &\leqslant 8.7, \\ 5x - 3y &\leqslant 10.0, \end{aligned}$$

Maximize $11x + 10y$.

5–6. Solve the following problem.

$$\begin{aligned} 2x_1 + x_2 + x_3 &= 10, \\ -44x_1 - 42x_2 + x_4 \qquad\qquad - x_8 &= -183, \\ 36x_1 - 102x_2 \qquad + x_5 \qquad - x_8 &= 17, \\ -164x_1 + 298x_2 \qquad\qquad + x_6 \quad - x_8 &= 1517, \\ -12x_1 - 6x_2 \qquad\qquad\qquad + x_7 - x_8 &= -79, \end{aligned}$$

$x_i \geqslant 0$ for $i = 1, 2, 3, 4, 5, 6, 7$.

Minimize x_8 (not sign-restricted !).

5–7. Solve the following problem by the Two-phase method.

Maximize $0.98n + 0.06x_1 + 0.15x_2 + 0.3x_3$,

where n is a constant, subject to

$$x_1 \leqslant n, \qquad x_2 - n_2 \leqslant 0, \qquad x_3 - n_3 \leqslant 0,$$
$$n_2 + 0.3x_1 = 0.6n, \qquad n_3 + 0.18x_1 + 0.3x_2 = 0.36n$$

(n_2 and n_3 are unknowns).

5–8. Solve the following problem.

$$x_1 - x_2 + x_3 - x_4 = 2,$$
$$2x_1 - 2x_2 - x_3 + x_4 = 1, \quad \text{All variables non-negative.}$$
$$4x_1 - 4x_2 + x_3 - x_4 = 5.$$

Minimize $x_1 + x_2 + x_3 + x_4$. (Note: One of the equations is redundant.)

5–9. Which constraints, and which objective function, are implied in the following tableau? Find the minimum of the objective function.

	x_1	x_2	x_3	x_4	x_5	x_6	
x_7	2	1	0	0	0	0	10
x_8	16	-12	-1	0	0	2	44
x_9	-12	34	0	-1	0	1	42
x_{10}	0	0	0	0	-1	1	0
	4	22	-1	-1	-1	4	86

5–10. Find all basic optimal solutions of

$$2x_1 + x_2 + z_1 = 6,$$
$$4x_1 + 2x_2 + z_2 = 12.$$

Minimize $z_1 + z_2$.

5–11. Solve the following problem (a) by the M-method and (b) by the Dual Simplex method.

$$y_3 + y_4 - 2y_5 - y_1 = 1,$$
$$y_3 - 2y_4 + y_5 - y_2 = -1.$$

Minimize $3y_3 + y_4 + 2y_5$.

5–12. (a) Find all optimal basic feasible solutions of

$$12x_1 + 7x_2 + 14x_3 \mid 5x_4 + 16x_5 \leqslant 1,$$
$$7x_1 + 14x_2 + 5x_3 + 16x_4 + 3x_5 \leqslant 1,$$
$$14x_1 + 5x_2 + 16x_3 + 3x_4 + 18x_5 \leqslant 1,$$
$$5x_1 + 16x_2 + 3x_3 + 18x_4 + x_5 \leqslant 1,$$
$$16x_1 + 3x_2 + 18x_3 + x_4 + 20x_5 \leqslant 1.$$

All x_i are non-negative.

Maximize $x_1 + x_2 + x_3 + x_4 + x_5$, using the Inverse Matrix method.

(b) Discuss the solutions to the dual problem.

(c) Discuss the problem with initial matrices M:

$$\begin{pmatrix} 1 & 12 & 7 & 14 & 5 & 1 & 0 & 0 & 0 & 0 \\ 1 & 7 & 14 & 5 & 16 & 0 & 1 & 0 & 0 & 0 \\ 1 & 14 & 5 & 16 & 3 & 0 & 0 & 1 & 0 & 0 \\ 1 & 5 & 16 & 3 & 18 & 0 & 0 & 0 & 1 & 0 \\ 0 & -1 & -1 & -1 & -1 & 0 & 0 & 0 & 0 & 1 \end{pmatrix},$$

$$\begin{pmatrix} 1 & 12 & 7 & 14 & 1 & 0 & 0 & 0 \\ 1 & 7 & 14 & 5 & 0 & 1 & 0 & 0 \\ 1 & 14 & 5 & 16 & 0 & 0 & 1 & 0 \\ 0 & -1 & -1 & -1 & 0 & 0 & 0 & 1 \end{pmatrix},$$

$$\begin{pmatrix} 1 & 12 & 7 & 1 & 0 & 0 \\ 1 & 7 & 14 & 0 & 1 & 0 \\ 0 & -1 & -1 & 0 & 0 & 1 \end{pmatrix}.$$

CHAPTER 6

SPECIAL ALGORITHMS

The algorithms described in Chapter 5 solve any linear programming problem, but we do not advocate their use in all cases. It would be a failure of ingenuity—and often of computer time—if we did not recognize and take advantage of special features which make the solution quicker and more efficient. In this chapter we shall justify special methods to deal with special cases, starting with possibly the simplest of them.

6-1 Transportation problem. The first problem we consider is the Transportation Problem, already mentioned in Chapter 1. The analytical formulation of its simplest form is

$$\sum_j x_{ij} = a_i, \qquad \sum_i x_{ij} = b_j.$$

$$\text{All } x_{ij} \text{ non-negative.}$$

$$\text{Minimize } \sum_i \sum_j c_{ij} x_{ij}.$$

When we add, separately, the equations of the two sets, we find that the two left-hand sides are equal, and therefore the two sets combined are only consistent if $\sum_i a_i = \sum_j b_j$. If this is the case, then only $m + n - 1$ of the $m + n$ equations are linearly independent, and hence a basic solution consists of $m + n - 1$ basic and of $mn - (m + n - 1) = (m - 1)(n - 1)$ nonbasic variables. The whole system is, of course, symmetric as to sources and destinations.

Explicitly, the constraints look as follows (taking the case 2 by 3 as an illustration):

$$
\begin{aligned}
x_{11} + x_{12} + x_{13} & & & = a_1, \\
& x_{21} + x_{22} + x_{23} & & = a_2, \\
x_{11} & + x_{21} & & = b_1, \\
x_{12} & + x_{22} & & = b_2, \\
x_{13} & + x_{23} & = b_3.
\end{aligned}
$$

The pattern of coefficients is very simple and, in particular, most of them are zero. For a more compact representation of the problem, we introduce two tables, the *requirement table*, and the *cost table*, as follows:

Requirement table · · · · · · Cost table

	b_1	b_2	b_m
a_1	x_{11}	x_{12}	x_{1m}
a_2	x_{21}	x_{22}	x_{2m}
.			
.			
.			
a_n	x_{n1}	x_{n2}	x_{nm}

c_{11}	c_{12}	c_{1m}
c_{21}	c_{22}	c_{2m}
.		
.		
.		
c_{n1}	c_{n2}	$c_{nm}.$

In the first, the totals of the variables in any row or column are given on the margin of that row or column. In the second, c_{ij} is the cost of one unit of the x_{ij} with the same subscripts.

The cost table remains unaltered during the algorithm; the requirement table contains at each stage the values of the basic variables.

Before we proceed, consider the constraints

$$\sum_j x_{ij} \leqslant a_i, \qquad \sum_i x_{ij} \leqslant b_j,$$

i.e. inequalities instead of the equations in the simplest case. In order to have again equations, we introduce a further set of non-negative variables y_i and z_j, and obtain the equivalent set of constraints:

$$\sum_j x_{ij} + y_i = a_i, \qquad \sum_i x_{ij} + z_j = b_j.$$

But now we have destroyed the special pattern of the earlier case, and in order to restore it, we add the following two equations:

$$\sum_i y_i + t = \sum_i a_i, \qquad \sum_j z_j + t = \sum_j b_j.$$

We identify t with $\sum_i \sum_j x_{ij}$, and then these two equations do not imply any new effective constraints. The requirement table which we have now reached is

	b_1	b_m	$\sum a_i$
a_1	x_{11}	x_{1m}	y_1
.			
.			
.			
a_n	x_{n1}	x_{nm}	y_n
$\sum b_j$	z_1	z_m	t

If only the first of the two sets of constraints consists of inequalities, while the second consists of equations, then we do not introduce the z_j, and instead of the last two equations we introduce just one, viz.

$$\sum_i y_i = \sum_i a_i - \sum_j b_j.$$

Again, this does not introduce any further effective constraint.

There exist, of course, feasible solutions to the transportation problem, e.g.

$$x_{ij} = \frac{a_i b_j}{\sum_i a_i} = \frac{a_i b_j}{\sum_j b_j}.$$

Hence a basic feasible solution must also exist. It can be constructed as follows: Take x_{11}, the variable in the *"northwest corner"* of the table, and make it equal either to a_1 or to b_1, whichever is smaller. Omit the line (row or column) of the latter, and subtract the value of x_{11} from the total of the other. We obtain then a smaller problem, which can be dealt with in the same way.

EXAMPLE 6–1.

	1	6	2	6
5	1			
5				
5				

	1	6	2	6
5	1	4		
5				
5				

	1	6	2	6
5	1	4		
5		2		
5				

	1	6	2	6
5	1	4		
5		2	2	
5				

	1	6	2	6
5	1	4		
5		2	2	1
5				

	1	6	2	6
5	1	4		
5		2	2	1
5				5

Thus we obtain a feasible solution, by setting the remaining variables equal to zero. In the present case the positive values number $m + n - 1$, but this would not be so if an entry before the last balanced a row and a column at the same time, as in the following example.

EXAMPLE 6–2.

	1	4	4	6
5	1	4		
5				
5				

At this stage, one row and two columns have been exhausted, and continuing with the remainder we obtain

	1	4	4	6
5	1	4		
5			4	1
5				5

We have now 4 positive values, less than $m + n - 1 = 5$. This is, of course, a case of degeneracy, and we must find a further basic variable, to which we give the value 0.

The simplest way of achieving this is to consider, at each entry, only either a row or a column, exhausted, but not both at the same time. We consider therefore that in one of them there is a remainder, zero in a degenerate case, and proceed by entering it in a next cell. Thus, in the case above, we would obtain either

	1	4	4	6
5	1	4	0	
5			4	1
5				5

or

	1	4	4	6
5	1	4		
5		0	4	1
5				5

where we have made the entries for basic variables.

A degenerate case is one where a partial sum of the row totals equals a partial sum of the column totals. It depends on the order in which we consider the rows and columns whether the difficulty inherent in having a row and a column simultaneously exhausted arises.

In Chapter 4 we pointed out the connection between a directed network and the Transportation Problem. Omitting the arcs from the master-source, and those to the master-sink, we obtain a graph whose incidence matrix is the matrix of coefficients of the transportation constraints, and a basis of the incidence matrix corresponds to a basic feasible solution. Hence the latter is represented by a tree in the graph.

It is now clear that even if we have $m + n - 1$ positive values in the cells of the requirement table, they need not describe a basic solution. If we draw an arc for each positive value, this may result in a disconnected graph with at least one loop in it. We have seen in Chapter 4 how to find out whether such a loop is present. If there are $m + n$ or more positive entries, then the associated graph must have at least one loop.

If the set of basic variables in a basic feasible solution is known, then we can find their values by modification of the northwest corner rule, and we see that these values must be integers. But if we have some nonbasic solution, then this is not necessarily so, as is seen from the following example:

		a	b	c	d
		1	4	4	6
A	5	1	$2\frac{1}{2}$	$1\frac{1}{2}$	
B	5		$1\frac{1}{2}$	2	$1\frac{1}{2}$
C	5			$\frac{1}{2}$	$4\frac{1}{2}$.

The graph which has arcs for those connections whose cells have a positive entry is now that in Fig. 6-1. There is no difficulty in spotting loops in it.

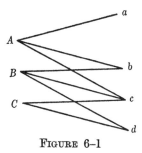

FIGURE 6-1

Although the northwest corner rule will always produce a basic feasible solution, it is advantageous to start as near to the optimum as possible, and this can be attempted by taking some account of the costs when determining a first basic feasible solution. We do this by starting from (one of) the cheapest cell(s), then filling in the next cheapest in the row or column which is not exhausted, and so on.

EXAMPLE 6-3.

Cost Table				Requirements				
					1	6	2	6
5	4	3	2	5				5
10	8	4	7	5	1	4		
9	9	8	4	5		2	2	1

The tie between the two cost entries 9 in the last row has been arbitrarily resolved.

It is also important to realize that the solution of a transportation problem remains unaltered if we add a positive or negative constant to all entries of the same row, or the same column. To prove this, assume that we subtract, from all costs $c_{i_0 j}$ of the i_0th row the constant c_{i_0}. The problem becomes then that of minimizing, subject to unaltered constraints, the objective function:

$$\sum_j (c_{i_0 j} - c_{i_0}) x_{i_0 j} + \sum_{i \neq i_0} \sum_j c_{ij} x_{ij}$$
$$= \sum_i \sum_j c_{ij} x_{ij} - c_{i_0} \sum_j x_{i_0 j} = \sum_i \sum_j c_{ij} x_{ij} - c_{i_0} a_{i_0}.$$

The last term is a constant, so that we are faced with minimizing the same double sum as before.

If $c_{i_0 j} - c_{i_0}$ is non-negative for all j, then $c_{i_0} a_{i_0}$ is a lower bound of the total cost.

EXAMPLE 6–3 (continued). We make use of this idea by subtracting the smallest cost in each row from the cost rates of that row, and then treating columns in a similar way. Starting from the cost table of the previous example, we subtract from the three rows, respectively, 2, 4, and 4, to obtain

3	2	1	0
6	4	0	3
5	5	4	0,

and then, subtracting from the first two rows 3 and 2 respectively

0	0	1	0
3	2	0	3
2	3	4	0.

We could also have started with the columns; then we would have obtained

0	0	0	0
5	4	1	5
4	5	5	2.

We have now discussed how to obtain a first feasible solution. It will have been noticed that in this respect the Transportation Problem is less difficult to deal with than the general Simplex method. We must now explain how we find out whether a basic feasible solution is optimal and, if it is not, how to improve on it. We shall be guided by our knowledge of the relationship of duality and of its consequences.

In writing the dual to the Transportation Problem, we write u_i for the variable corresponding to the ith equation of the first set, and v_j for that of the jth equation in the second set. These variables are not sign-restricted, because the constraints in the primal set are equations. The dual system is then

$$u_i + v_j \leqslant c_{ij} \qquad \text{(all } i, j\text{)}.$$
$$\text{Maximize } \sum_i a_i u_i + \sum_j b_j v_j.$$

Assume that we have found $m + n - 1$ basic variables x_{ij}. Then the corresponding $m + n - 1$ dual relations, i.e. those with the same (i, j) pairs, can be satisfied as equations $u_i + v_j = c_{ij}$. Because the arcs from the ith source to the jth destination in the graph of our problem form a tree, these equations are triangular and can be solved in a trivial way; because there are $m + n$ unknowns, but only $m + n - 1$ independent equations, one of the unknowns can take an arbitrary value, say $u_1 = 0$.

EXAMPLE 6–3 (continued).

	1	6	2	6
5	1	4		
5		2	2	1
5				5

The basic variables are x_{11}, x_{12}, x_{22}, x_{23}, x_{24}, and x_{34}. To find the values of the dual variables, we have to solve

$$u_1 + v_1 = 5, \qquad u_2 + v_2 = 8, \qquad u_2 + v_4 = 7,$$
$$u_1 + v_2 = 4, \qquad u_2 + v_3 = 4, \qquad u_3 + v_4 = 4.$$

This is the system considered in Example 4–6 (though with a different notation). We obtain

u_1	u_2	u_3		v_1	v_2	v_3	v_4
0	4	1		5	4	0	3 .

The variables u_i and v_j are called *shadow costs* (of. Section 7–1) and are tabulated as follows:

	1	6	2	6	Shadow costs
5	1	4			0
5		2	2	1	4
5				5	1
Shadow costs	5	4	0	3 .	

If they satisfied $u_i + v_j \leqslant c_{ij}$ for all pairs (i, j), the final answer would have been reached. In the present case this is not so. For instance, $u_1 + v_4 = 3 > c_{14} = 2$. Hence the present u_i and v_j are not part of a feasible solution. At least one of the slack variables has a negative value. Using the Dual Simplex method, we can make it non-basic—for the primal this means that the variable associated with the inequality in which that slack variable appeared will become basic.

Add the corresponding arc to the tree of the previous solution. It ceases to be a tree, but we obtain another tree again by dropping one of the arcs. Which one?

In terms of the tabular presentation, the loop which is contained in the enlarged graph corresponds to a succession of cells. Attach a plus sign to the cell of the new basic variable. Then assign, alternately, minus and plus signs to the successive cells corresponding to the successive arcs of the loop. These signs indicate additions and subtractions to be made to and from entries in order to balance the rows and columns. Moving from one of the signed cells to the next we move, alternately, along horizontal and vertical lines.

We transfer as many units as is possible without making any entry negative, i.e. as many as the smallest entry marked with a minus sign indicates. Thereby one cell previously filled in will become empty, and we have reached a new basic feasible solution.

In the dual problem, the value of the objective function has been reduced (remember that we used the Dual Simplex method!) and so has, therefore, also the total of the cost in the primal, except for degenerate cases, where it might have remained unaltered. This can also be seen, without reference to duality, as follows. Let the cell to be filled in be defined by (i_0, j_0), and let those taking part in the transfer be

$$(i_0, j_0), \ (i_0, j_1), \ (i_1, j_1), \ . \ . \ . \ ., \ (i_k, j_k), \ (i_k, j_0),$$

returning thereafter to (i_0, j_0). Because all these cells except the first belong to basic variables, we have for all these $u_i + v_j = c_{ij}$. The transfer results therefore in a change of the total cost amounting to

$$+c_{i_0 j_0} - u_{i_0} + u_{i_1} - u_{i_1} \ . \ . \ . + u_{i_k} - u_{i_k}$$
$$- v_{j_1} + v_{j_1} - v_{j_2} \ . \ . \ . + v_{j_k} - v_{j_0}$$
$$= c_{i_0 j_0} - u_{i_0} - v_{j_0}.$$

Hence, when

$$c_{i_0 j_0} < u_{i_0} + v_{j_0},$$

then the change results in a saving.

Any solution with more than $m + n - 1$ pairs (i, j) for which $u_i + v_j = c_{ij}$, leads to a loop in the graph where the corresponding arcs are drawn. Then we can drop one of the arcs, and transfer the entry to the cells corresponding to the other arcs of the loop, without altering the total cost. It follows that nothing can be gained by considering other than basic feasible solutions.

If $u_i + v_j = c_{ij}$ for some cell which has not been entered, then a transfer can be made into it without changing the total cost. If the solution is optimal, then such a procedure produces alternative optimal solutions.

EXAMPLE 6–3 (continued). We know that we could fill in the cell (1, 4) with advantage and the plus and minus signs indicate what transfers this involves:

	1	6	2	6
5	1	4⁻		+
5		2⁺	2	1⁻
5				5 .

We cannot transfer more than 1, because otherwise the entry in the cell (2, 4) would become negative. Hence we obtain

	1	6	2	6	Shadow costs
5	1	3		1	0
5		3	2		4
5				5	2
Shadow costs	5	4	0	2	

The new shadow costs should be obtained from the triangular equations which they satisfy, or also from the graph.

We have changed (cf. Example 4–6) the graph of Fig. 6–2(a) into

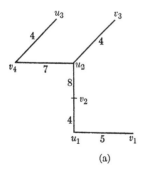

(a)   (b)

FIGURE 6–2

that of part (b). Thus only v_4 and u_3 are affected. Only these are now reached through a different path from u_1. The total cost is 71.

The comparison of the shadow costs with the true costs shows that we have now reached the optimum.

This can again be confirmed independently of considerations of duality. We know that the problem does not change if we subtract constants from the rows and columns. Subtract, then, the u_i from the respective rows, and the v_j from the columns. We obtain

$$
\begin{array}{cccc}
0 & 0 & 3 & 0 \\
1 & 0 & 0 & 1 \\
2 & 3 & 6 & 0.
\end{array}
$$

In the last solution only cells with zero cost were used. Since all other cells have positive cost, using any of these would necessarily increase the total cost.

In our method of solving the Transportation Problem we have never made use of division, and this shows once more (cf. Section 4–8) that if the marginal totals (the a_i and b_j) are integers, the optimal answer will also be expressed in integers. Of course, if there exist alternative basic optimal solutions, any linear combination with positive fractional coefficients will also be an optimal solution, but not necessarily in integers.

We have seen how to find a first basic feasible solution in a degenerate case as well. But it might then happen that the smallest amount with a minus sign in the preparation for an improving transfer is 0, so that the transfer thereby indicated does not change the total cost. We are then again in danger of getting into a loop and never reaching the optimum, or perhaps not recognizing it when we have reached it. Such a case can again be resolved by the application of a perturbation method.

A perturbation takes account of the fact that a case is degenerate when some partial sum of the a_i equals some partial sum of the b_j. This can be repaired by adding to the row and the column totals appropriate multiples of a small positive amount, which will be ignored in the final answer. The multiples must be judiciously chosen, to be certain that no partial sums remain equal; also, for convenience in automatic computation, where the small additions must have numerical values, we want their disappearance to be equivalent to rounding the fractional values to the nearest integers (see [107]).

Let $m \geqslant n$, and let d be the value of the unit in the least significant digit. For instance, if we know the answers to be integers, $d = 1$. The first objective is achieved by making the small addition, x, say, smaller than d/n, changing all a_i into $a_i + x$ $(i = 1, \ldots, n)$, b_1 into $b_1 + nx$, and leaving all other b_j unaltered. It is then clear that no equality of partial sums can occur.

However, to achieve the second objective, we make $x < d/2n$, which satisfies the first condition as well. Every entry x_{ij} in the requirement table is the difference between two partial sums (where 0 is also considered to be such a partial sum), say $\sum a_i$ and $\sum b_j$. With the additions of the multiples of x such a difference will be $|\sum_i a_i - \sum_j b_j| + tx$, where t is easily seen to be such that $|t| \leqslant n$, and hence $|tx| < \frac{1}{2}d$. It follows that rounding to the nearest integral multiple of d, we obtain precisely

$$|\sum_i a_i - \sum_j b_j| = x_{ij}.$$

The x_{ij} thus found optimize the original problem, because the shadow costs depend only on the pairs (i, j) of the basic variables, so that the criteria for optimality of the modified and of that of the original problem are simultaneously satisfied.

EXAMPLE 6–4.

Cost table					Requirement table

						1	4	4	6
5	4	3	2		5				
10	8	4	7		5				
9	9	8	4		5				

To avoid degeneracy, we choose $x < 1/6$, say $x = 0.1$, and have the following succession of solutions:

	1.3	4	4	6	Shadow costs
5.1	1.3	3.8⁻	+		0
5.1		0.2⁺	4	0.9⁻	4
5.1				5.1	1
Shadow costs	5	4	0	3	

	1.3	4	4	6	Shadow costs
5.1	1.3	2.9		0.9	0
5.1		1.1	4		4
5.1				5.1	2
Shadow costs	5	4	0	2	

The second table is optimal, and indicates that the original problem is finally solved by

	1	4	4	6
5	1	3		1
5		1	4	
5				5 .

6–2 Transshipment. It is conceivable that the route from a port of departure to a port of destination is shorter if intermediate calls are made at some other port or ports, either of departure, or of destination, and we discuss such a case by means of an example.

EXAMPLE 6–5. Denote sources by S_i and destinations by D_j, and let the cost table be as follows:

	D_1	D_2	D_3	D_4	S_1	S_2	S_3
S_1	5	4	3	2	0	2	1
S_2	10	8	4	7	1	0	4
S_3	9	9	8	4	3	2	0
D_1	0	1	3	2	5	9	9
D_2	3	0	2	3	4	6	7
D_3	2	3	0	1	3	4	9
D_4	4	1	2	0	4	7	3 .

Let the requirements and availabilities be the same as in Example 6–3. The cost table is not symmetrical. This is realistic, for instance, for road transport with one-way streets. One could also think of applications where the diagonal cells (i.e. those in cells where the row and column have the same label) are not all zero. The top left 3×4 table is identical with that in Example 6–3. Notice that it is quicker to sail from S_2 to D_2 via S_1 and D_4, for instance, and there are other transshipment routes which are cheaper than the direct route.

We might be tempted to deal with this case in the usual transportation manner. However, if the required row and column totals are those which have been used before, then the units available at the D_j, and also those required at the S_i, are zero, and hence no transshipment would ever arise. We overcome this difficult by adding to all row totals as well as to all column totals a large amount, L say, large enough to cover any possible transshipment, so that the optimal answer will still have positive values in all diagonal cells. L is certainly large enough if it equals the sum of all initial availabilities, or all initial requirements.

In the present example this is 15, and for simplicity of writing we choose $L = 100$. (We could equally well make all column totals L, and use $L + a_i$ and $L - b_j$ as row totals.)

We start from the solution of the example without transshipment, adding 100 in all diagonal cells:

	101	106	102	106	100	100	100	Shadow costs
105	1	3+		1	100−			0
105		3−	2		+	100		4
105				5			100	2
100	100							−5
100		100						−4
100			100					0
100				100				−2
Shadow costs	5	4	0	2	0	−4	−2	

A final solution (verify from the shadow costs!) is as follows:

	101	106	102	106	100	100	100	Shadow costs
105	1			7	97			0
105			2		3	100		1
105				5			100	2
100	100							−5
100		100						−3
100			100					−3
100		6		94				−2
Shadow costs	5	3	3	2	0	−1	−2	

In order to extract from this the optimal transportation schedule, we omit the entries in diagonal cells and obtain

	D_1	D_2	D_3	D_4	S_1
S_1	1			7	
S_2			2		3
S_3				5	
D_4		6			

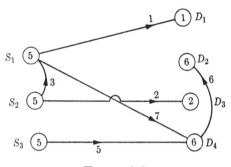

FIGURE 6-3

The total cost of this scheme is 56 (Fig. 6-3).

If we have originally m sources and n destinations, then the extended problem has $m + n$ sources, and the same number of destinations. A basic solution contains now $2n + 2m - 1$ basic variables, and omitting those in the $m + n$ diagonal cells we are left with $m + n - 1$ basic variables again.

We have thus shown that we obtain a basic feasible solution for the original transshipment problem, but have not yet shown that it is optimal. We shall now prove that this is so, as a consequence of the fact that there are positive entries in the diagonal cells of the final solution to the problem with extended marginal sums. Our presentation is based on [108].

We assume now that the costs c_{ij} are positive for $i \neq j$ and that all $c_{ii} = 0$.

The minimum cost of the complete transshipment problem depends on L in such a way that for increasing L the minimum total cost remains either constant or decreases. It cannot increase, because if the minimum cost for $L = L_1$, say, has been obtained, and if then the diagonal elements are increased by $L_2 - L_1 > 0$, then we obtain a feasible solution for $L = L_2$, while the total cost remained the same (the diagonal cells have zero cost).

We want to show that there is a value of L such that after further increasing it the minimum total cost remains unchanged. But before we do that we show that for sufficiently large L all diagonal cells will receive positive entries in the optimal solution. Indeed, when there is a diagonal cell whose entry, x_{ii}, say, is zero, then the minimum total cost for a given L is larger than $c_0 L$, where c_0 is a value smaller than any c_{ij} for $i \neq j$ (since at least L is required in any row and in any column). But as L increases, $c_0 L$ increases without bounds, while this is impossible for the minimal cost, because this cost is finite for some

L_1, and not larger for larger L. Hence, for sufficiently large L, no x_{ii} can be zero.

Let now $\bar{L}$ be such a sufficiently large value. If for this value the optimal solution of the complete transshipment problem is given by $x_{ij}(\bar{L})$, say, then a solution for $L > \bar{L}$ is given by

$$x_{ij}(L) = x_{ij}(\bar{L}) \qquad \text{when } i \neq j,$$

and

$$x_{ii}(L) = x_{ii}(\bar{L}) + (L - \bar{L}).$$

The cost of the $x_{ij}(L)$ is the same as that of the $x_{ij}(\bar{L})$, because the costs on the diagonal cells are zero. But $x_{ij}(L)$ is also the optimal solution of the complete transshipment problem with $\bar{L}$, because the shadow costs of the two solutions satisfy the same conditions $u_i + v_j \leqslant c_{ij}$, and indicate thereby that the minimum has been found in both cases.

The transshipment argument is equally valid when there is only one source, and one destination, and intermediate stations which are neither. The problem of the shortest path between two towns, via intermediate road junctions, which has been dealt with in Section 4–2, belongs within this category. In Example 4–1 we have an availability of 1 in M, and a requirement of 1 in H, with all other towns as intermediate stations. (See Exercise 6–4.)

6–3 Transportation with capacity restrictions. In many practical applications it is realistic to assume that the amount which can be sent on any particular route is restricted by the capacity of that route. Also, when a route is altogether excluded, this can be expressed by limiting its capacity to zero; this is an alternative to attaching to that route a very high cost.

When the cost of a particular route is unknown, then it may be of interest to find a minimal cost schedule without using that route, and the shadow costs will then allow us to determine that limit of the unknown cost above which the route would not be a part of any best schedule.

The constraints for a *capacitated* Transportation Problem are

$$\sum x_{ij} = a_i \quad \text{(all } i\text{)}, \qquad \sum_i x_{ij} = b_j \quad \text{(all } j\text{)},$$

and

$$x_{ij} \leqslant r_{ij} \qquad \text{(all } i, j\text{)},$$

where r_{ij} is the capacity limit on route (i, j). The objective function is again $\sum_i \sum_j c_{ij} x_{ij}$, where the c_{ij} are unit costs.

Such a problem does not always have a solution. For instance, if a marginal total is larger than the sum of the capacities of all cells in that row or column, then the conditions can, clearly, not be satisfied. If the limitations consist merely of the exclusion of some routes, then using a very high cost for these will indicate the inconsistency of the conditions by leaving one of those cells with a positive entry in the optimal solution.

There exist various ways of transforming a capacitated Transportation Problem into one of the ordinary type, and we mention two of them. These two are not recommended for computational purposes (Exercise 6–10 will give an idea of their efficiency), but they are interesting, because they show that if the a_i, b_j, and r_{ij} are integers, then the capacitated problem has optimal solutions in integers.

H. M. Wagner [125] deals with the above problem as follows:

Requirement table

	b_1	b_2	...	b_m	$\sum_j r_{1j} - a_1$	$\sum_j r_{2j} - a_2$	...	$\sum_j r_{nj} - a_n$
r_{11}	x_{11}				y_{11}			
$\cdot$								
$\cdot$								
$\cdot$								
r_{1m}				x_{1m}	y_{1m}			
r_{21}	x_{21}					y_{21}		
$\cdot$								
$\cdot$								
$\cdot$								
r_{2m}				x_{2m}		y_{2m}		
$\cdot$								
$\cdot$								
$\cdot$								
r_{n1}	x_{n1}							y_{n1}
$\cdot$								
$\cdot$								
$\cdot$								
r_{nm}				x_{nm}				y_{nm}

The cost of x_{ij} is c_{ij}, that of y_{ij} is zero, and that of the cells without a variable in the table above is M, say, i.e. very high. The row sums are $x_{ij} + y_{ij} = r_{ij}$, and the column sums are in the first portion

$\sum_i x_{ij} = b_j$, and in the second $\sum_j y_{ij} = \sum_j r_{ij} - a_i$ which reduces, from the row sums, to $\sum_j x_{ij} = a_i$.

Another transformation was introduced by G. B. Dantzig. His requirement table, written for $n = 2$, $m = 3$, looks as follows (in our present notation):

	r_{11}	r_{12}	r_{13}	r_{21}	r_{22}	r_{23}	b_1	b_2	b_3
a_1	x_{11}	x_{12}	x_{13}						
r_{11}	y_{11}						x_{11}		
r_{12}		y_{12}						x_{12}	
r_{13}			y_{13}						x_{13}
a_2				x_{21}	x_{22}	x_{23}			
r_{21}				y_{21}			x_{21}		
r_{22}					y_{22}			x_{22}	
r_{23}						y_{23}			x_{23}

Two cells containing the same variable x_{ij} will have the same value by virtue of the repeated marginal totals r_{ij}, without any precautions being taken to that effect.

A method mentioned by A. Charnes and W. W. Cooper in [35] makes again use of the relationship between the values of the variables of optimal solutions in dual problems. Consider a capacitated Transportation Problem and its dual

$$\sum_j x_{ij} = a_i, \qquad\qquad u_i + v_j + w_{ij} \leqslant c_{ij} \quad \text{all } i, j$$
$$\text{(corresponding to } x_{ij}\text{),}$$

$$\sum_i x_{ij} = b_j, \qquad\qquad w_{ij} \leqslant 0 \quad \text{all } i, j$$
$$\text{(corresponding to } y_{ij}\text{).}$$

$$x_{ij} + y_{ij} = r_{ij}.$$

Minimize $\sum_i \sum_j c_{ij} x_{ij}$.   Maximize $\sum_i a_i u_i + \sum_j b_j v_j + \sum_i \sum_j r_{ij} w_{ij}$.

We try first to find a feasible solution of the primal problem. If

$$x_{ij} > 0 \quad \text{and} \quad y_{ij} > 0 \quad \text{(i.e. } x_{ij} < r_{ij}\text{),}$$

then we must have

$$u_i + v_j = c_{ij}, \qquad w_{ij} = 0 \quad \text{for the same pair } i, j.$$

If

$$x_{ij} > 0 \quad \text{and} \quad y_{ij} = 0 \quad \text{(i.e. } x_{ij} \text{ has reached its upper limit),}$$

then we must have

$$u_i + v_j + w_{ij} = 0, \qquad w_{ij} \leqslant 0$$

or, in other words,

$$u_i + v_j \geqslant 0 \qquad \text{for the same } i, j,$$

and, finally, if

$$x_{ij} = 0 \quad \text{and hence} \quad y_{ij} > 0,$$

then we must have

$$u_i + v_j \leqslant 0, \qquad w_{ij} = 0.$$

The only new feature is the requirement that if a variable reaches its upper bound, then $u_i + v_j < 0$ indicates that a reduction is desirable. The amount of transfers equals either the smallest flow from which a reduction is to be made, or the smallest residual in cells where increases are indicated, whichever of these two is smaller. The u_i and v_j must be found from cells which form a basic solution for the problem without capacity restrictions.

EXAMPLE 6–6.

	3	5	5	Shadow costs		Cost table		
	1)	1)	2)	0		6	1	1
2		1^{-}	1^{+}			5	2	7
	2)	1)	2)	6		9	7	8
2			2					
	3)	5)	4)	7				
9	3	4^{+}	2^{-}					

Shadow costs   2   0   1

	3	5	5	Shadow costs
	1)	1)	2)	
2			2	0
	2)	1)	2)	
2	+		2^{-}	6
	3)	5)	4)	
9	3^{-}	5	1^{+}	7

Shadow costs   2   0   1

	3	5	5	Shadow costs
	1)	1)	2)	
2			2	0
	2)	1)	2)	
2	2^{-}	+		3
	3)	5)	4)	
9	1^{+}	5^{-}	3	7

Shadow costs   2   0   1

6-4] NETWORK FLOW METHOD 135

	3	5	5	Shadow costs
	1)	1)	2)	
2			2	0
	2)	1)	2)	
2	1	1		3
	3)	5)	4)	
9	2	4	3	7
Shadow costs	2	0	1	

It is worth comparing this with the result of Example 4–5, which used the same network. The present solution implies a maximal flow alternative to that found there. In the present case we have a unique solution, due to the consideration of cost as well.

A basic solution to a capacitated Transportation Problem may very well contain more than $m + n - 1$ positive values, due to the capacity constraints which are additional to the $m + n - 1$ independent equations in the simpler case.

If the conditions of a capacitated Transportation Problem are inconsistent, then we might still be interested in knowing the maximal possible network flow from $\oplus$ to $\ominus$. If they are consistent, then the problem can be expressed as requiring a maximal flow (which will be equal to $\sum_i a_i = \sum_j b_j$), and of all alternative flows we want to know that one in particular which is cheapest.

This formulation involves two objective functions, one of them being overriding. We can express this by using a device which may be useful in other contexts as well. We choose as the objective function to be maximized

$$\sum_j x_{\oplus j} - \varepsilon \sum_i \sum_j c_{ij} x_{ij},$$

where ε is so small that it can be ignored in comparison with any other amount. The double sum has then only an effect if there are multiple maxima to the first sum—as there will always be in an ordinary Transportation Problem with more than one source and one destination.

6-4 Network flow method. We turn now to the Ford–Fulkerson method of solving the Transportation Problem, deriving it from the *Primal-Dual algorithm* discused in Section 5–11. Consider the problem

$$\sum_j x_{ij} = a_i, \qquad \sum_i x_{ij} = b_j, \qquad x_{ij} \geqslant 0.$$
$$\text{Minimize } \sum_i \sum_j c_{ij} x_{ij}.$$

Its dual is

$$u_i + v_j \leqslant c_{ij}, \qquad \text{for all } i, j.$$

$$\text{Maximize } \sum_i u_i a_i + \sum_j b_j v_j.$$

If, for a start, we subtract in each row the smallest cost from all cost entries, then

$$u_i = 0, \qquad v_j = 0, \qquad \text{for all } (i, j)$$

is certainly a solution of the dual problem.

Let the set of all those j which indicate cells with $u_i + v_j = c_{ij}$ for a given i be denoted by S_i, and the set of all i for which $u_i + v_j = c_{ij}$ for a given j by T_j. Then the restricted primal is

$$\sum_{j \text{ in } S_i} x_{ij} + s_i = a_i \qquad (\text{all } i),$$

$$\sum_{i \text{ in } T_j} x_{ij} + t_j = b_j \qquad (\text{all } j).$$

$$\text{Minimize } \sum_i s_i + \sum_j t_j.$$

The form of the constraints shows that minimizing the objective function is equivalent to maximizing the flow through those cells (i, j) for which $u_i + v_j = c_{ij}$. This flow is

$$\sum_i \sum_{S_i} x_{ij} = \sum_j \sum_{T_j} x_{ij}.$$

The problem of the maximal network flow, which is here that of solving the restricted primal, can be solved by the labelling method explained in Chapter 4.

EXAMPLE 6–7. We take again the case of Example 6–3.

Cost table				Requirements				
					1	6	2	6
5	4	3	$\underline{2}$	5				O
10	8	4	7	5		O		
9	9	8	$\underline{4}$	5				O

We have underlined the smallest cost in each row, and ringed the corresponding cells in the requirement table. We want to send the largest flow through the network, using only those arcs which correspond to ringed cells (Fig. 6–4).

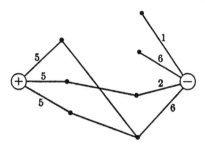

FIGURE 6–4

A maximal flow is clearly the following (if we did not see it at once, the labels would show it, because they do not lead to a column with a deficiency; a and b have deficiencies, but cannot be labelled):

	a	b	c	d	
	1	6	2	6	
A 5				⑤	$(4, d)$
B 5			②		$(2, \oplus)$
C 5				①	$(4, \oplus)$
		$(3, B)$	$(4, C)$		

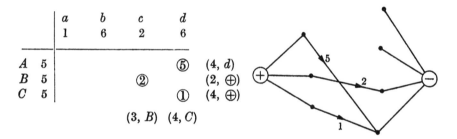

This concludes the solution of the restricted primal. Now we must find the solution to the dual of the latter, which can be written

$$g_i \leqslant 1 \quad \text{(all } i\text{)}, \qquad\qquad h_j \leqslant 1 \quad \text{(all } j\text{)},$$
$$g_i + h_j \leqslant 0 \quad \text{for all pairs } i \text{ which belong to ringed cells.}$$
$$\text{Maximize } \sum a_i g_i + \sum b_j h_j.$$

The solution can be found from that of the restricted primal, by taking those constraints to be equations which correspond to the basic variables in that primal. Such variables will be s_i (if there is a deficiency in the ith row), or t_j (if there is a deficiency in the jth column), or finally x_{ij} if there is a flow through the arc from S_i to D_j. The first leads to $g_i = 1$, the second to $h_j = 1$, and the third to $g_i + h_j = 0$.

EXAMPLE 6–7 (continued). The first solution to the dual was

$$u_1^0 = 2, \qquad u_2^0 = 4, \qquad u_3^0 = 4, \qquad v_j^0 = 0 \quad (j = 1, 2, 3, 4),$$

and hence the restricted primal is

$$x_{14} + s_1 = 5, \qquad x_{23} + t_3 = 2,$$
$$x_{23} + s_2 = 5, \qquad x_{14} + x_{34} + t_4 = 6.$$
$$x_{34} + s_3 = 5,$$

$$\text{Minimize } s_1 + s_2 + s_3 + t_3 + t_4.$$

The answer is (see the last requirement table)

$$x_{14} = 5, \qquad x_{23} = 2, \qquad x_{34} = 1,$$
$$s_1 = 0, \qquad s_2 = 3, \qquad s_3 = 4, \qquad t_3 = 0, \qquad t_4 = 0.$$

The dual to the restricted primal is

$$g_1 \leqslant 1, \qquad h_j \leqslant 1 \quad (j = 3, 4),$$
$$g_1 + h_4 = 0, \qquad g_2 + h_3 = 0, \qquad g_3 + h_4 = 0,$$
$$g_2 = 1, \qquad g_3 = 1.$$

$$\text{Maximize } 5(g_1 + g_2 + g_3) + 2h_3 + 6h_4.$$

The solution is

$$g_i^0 = 1 \quad \text{for } i = 1, 2, 3,$$
$$h_3^0 = h_4^0 = -1.$$

(The value of the objective function is 7.)

In general, g_i for a labelled row i will be 1, and h_j for a labelled column -1. The improved solution for the original dual is found by determining a value t as large as possible, and such that $u_i^0 + tg_i^0$ and $v_j^0 + th_j^0$ remain solutions. This is easily seen to mean that t is the smallest of $c_{ij} - u_i^0 - v_j^0$ for all labelled i and unlabelled j. The improved values of u_i and v_j are obtained by adding this value of t to all u_i for labelled rows, and subtracting it from all v_j for labelled columns. In all cases where $u_i + v_j$ equalled c_{ij} this equality holds also for $(u_i + t) + (v_j - t)$, and there will be at least one more equation in the dual constraints.

We thus continue the procedure of the Primal-Dual algorithm.

EXAMPLE 6–7 (continued). We have to take the smallest of

$$5 - 2 \qquad 4 - 2$$
$$10 - 4 \qquad 8 - 4$$
$$9 - 4 \qquad 9 - 4,$$

i.e. $t = 2$. This gives new shadow costs and ringed cells as follows:

				Shadow costs		1	6	2	6	
5	<u>4</u>	3	<u>2</u>	4	5	O			⑤	$(4, d)$
10	8	<u>4</u>	7	6	5			②		$(3, \oplus)$
9	9	8	<u>4</u>	6	5				①	$(4, \oplus)$.

Shadow costs 0   0 −2 −2 $(4, A)\ (3, B)\ (4, C)$

This leads to

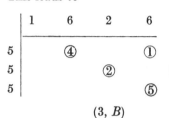

	1	6	2	6	
5		④		①	
5			②		$(3, \oplus)$
5				⑤	
		$(3, B)$			

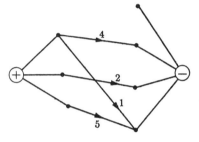

The improved solution to the dual is now found by determining the smallest of

$$10 - 6, \qquad 8 - 6, \qquad 7 - 4,$$

so that $t = 2$. Adding to the shadow cost of the labelled row, and subtracting from that of the labelled column we obtain

						1	6	2	6	
5	4	3	<u>2</u>	4	5	④			①	
10	<u>8</u>	<u>4</u>	7	8	5	O	②			$(3, \oplus)$
9	9	8	<u>4</u>	6	5				⑤	

0   0 −4 −2 $(3, B)\ (3, B)$

which leads to

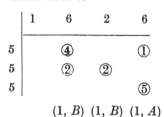

	1	6	2	6	
5		④		①	$(1, b)$
5		②	②		$(1, \oplus)$
5				⑤	$(1, d)$.
	$(1, B)$	$(1, B)$	$(1, A)$		

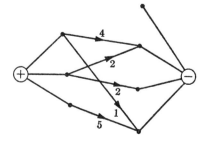

This is again the maximal flow under the given circumstances, so we find the smallest of

$$5 - 4,$$
$$10 - 8,$$
$$9 - 6,$$

i.e. 1 and add it to all u_i, and subtract it from v_2, v_3, and v_4:

						1	6	2	6	
$\underline{5}$	$\underline{4}$	3	$\underline{2}$	5	5	O	④		①	$(1, b)$
10	$\underline{8}$	4	7	9	5		②	②		$(1, \oplus)$
9	9	8	$\underline{4}$	7	5				⑤	

| 0 | -1 | -5 | -3 | | | $(1, A)$ | $(1, B)$ | $(1, B)$ |

leading to the final answer

	1	6	2	6
5	①	③		①
5		③	②	
5			⑤	.

There is no deficiency left, and the result is the same as that of Example 6–3 (Fig. 6–5).

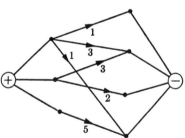

FIGURE 6–5

This method of solving the transportation problem is due to L. R. Ford and D. R. Fulkerson (see, in particular, [66]). It is also applicable to a capacitated problem. That part of the algorithm which deals with finding a maximal flow has been described in Chapter 4. The rule for changing the shadow costs, when a maximal flow has been reached and the next stage is entered, needs some modification. This arises from the fact that now the dual problem contains variables in addition to those interpreted as shadow costs. We refer the reader to [67].

6–5 The assignment problem. Consider the following problem. There are n jobs and n candidates, and it is required to allocate the candidates to the jobs in some optimal way. To this end the candidates are interviewed and the ith candidate obtains a mark r_{ij} indicating his fitness for the jth job. The assignment should be done in such a way that the total of the marks of the candidates in those jobs to which they are being assigned is as large as possible.

Let $x_{ij} = 1$ if the ith candidate is assigned to the jth job, and 0 otherwise. We have to solve the following problem:

$$\sum_i x_{ij} = 1 \quad (\text{all } j), \qquad \sum_j x_{ij} = 1 \quad (\text{all } i), \qquad x_{ij} \geqslant 0 \quad (\text{all } i, j).$$

$$\text{Maximize } \sum_i \sum_j r_{ij} x_{ij}.$$

This is a Transportation Problem, and therefore the values of the variables will be 0 or 1. It might be called the most degenerate Transportation Problem, because after every single allocation we exhaust a row and a column simultaneously. Such a problem could be solved by the perturbation method, but the Primal-Dual algorithm, or rather its adaptation to the Transportation Problem, is insensitive to degeneracy, and thus the assignment problem can be solved by it just as easily as a nondegenerate problem. We have to remember, of course, that in this case we *maximize* the objective function.

Since any change of allocation involves only one single person in any job, it is unnecessary to carry the number of persons to be reallocated along in the labels, and we shall therefore only record the sources (persons) or destinations (jobs). The changes contingent on maximizing rather than minimizing are trivial and easily understood. We shall mention them during the solution of the following example.

EXAMPLE 6–8. Let the rating marks be as follows:

		Job 1	Job 2	Job 3	Job 4	Shadow cost	Label
Person	1	④*	2	0	3	4	
	2	2	④*	④	3	4	(2)
	3	2	⑥	2	3	6	(0)
	4	4	⑤	3	4	5	
Shadow cost		0	0	0	0		
Label			(3)	(2)			

This time we have ringed the row *maxima* (because we want to maximize). We call cells *independent* if there are not two of them in any one row or in any one column. We have started by marking some such ringed cells with an asterisk.

The labels lead to

④*	2	0	3	4	
2	④	④*	3	4	
2	⑥*	2	3	6	(2)
4	⑤	3	4	5	(0)
0	0	0	0		
	(4)				

This is the maximal possible assignment, given the ringed cells. To find an improved solution to the dual (in the spirit of the Primal-Dual algorithm) we have to find the smallest of the differences $u_i + v_j - r_{ij}$ in labelled rows and unlabelled columns:

$$6 - 2 \qquad\qquad 6 - 2 \quad 6 - 3$$
$$5 - 4 \qquad\qquad 5 - 3 \quad 5 - 4.$$

Subtract 1 from the shadow costs of labelled rows and *add* 1 to those of labelled columns:

④*	2	0	3	4
2	4	④*	3	4
2	⑥*	2	3	5
④	⑤	3	④*	4
0	1	0	0	

It is obvious that we can now mark one of the newly ringed cells with an asterisk and obtain thereby the optimal assignment. The total of all marked r_{ij} is 18, and so is the total of all shadow costs, as it must be by the properties of duality.

6–6 The assignment problem. Hungarian method. When solving Example 6–8 we have applied the rules of the network flow method with its labels. It derives directly from the Primal-Dual algorithm when the problem is not degenerate. However, the present problem is degenerate, and consequently the dual to the restricted primal has more than one solution. Therefore an alternative procedure is possible for the improvement of the solution to the original dual problem.

The network flow method is thus not the only Primal-Dual algorithm that can be devised and we shall now introduce another method. It differs from the previous one on two accounts. It uses a method for solving the restricted primal, i.e. for finding the maximal flow or, equivalently, the maximum number of independent cells, which differs from the earlier one, and the values of t, leading to new ringed cells, and to new shadow costs, are also, in general, different from those determined earlier. The justification for both these steps is dependent on theorems due to Hungarian mathematicians, König and Egerváry, and the method is hence called Hungarian. It is due to H. W. Kuhn (see [96]).

We start with the same solution to the dual as in the network flow method, and we mark a number of independent cells by taking the first in the first row (if any), then the first in the second row which is not in a column with a marked cell, and so on.

After this, we have to see whether the largest possible number of independent cells has been marked. We know from König's theorem (Section 4–9) that this number is also the smallest number of lines which contain between them all ringed cells.

Our argument will be as follows: We introduce a routine for marking a line (row or column) through each assigned cell, i.e. one marked with an asterisk. If the lines drawn according to this routine do not pass through all ringed cells, then we can mark at least one more ringed cell, and draw more lines, as will be shown. Thus we proceed until all ringed cells are crossed by precisely as many lines as there are marked cells.

There will always be as many lines drawn as there are marked cells, and when all ringed cells are crossed by such lines, then the minimum number of crossing lines has been determined. By König's theorem the maximum number of independent ringed cells has then been marked.

To explain the working of the routine in detail, we introduce the concept of a *transfer*. By this we understand a change of assignment, where we start with an assigned person and reassign him, if possible, to an as yet unassigned eligible job (i.e. one indicated by a ringed cell). Let then the first person's job be taken by somebody else who was assigned to another job, and so on until eventually a job that was assigned remains now unassigned.

We investigate then for each marked cell whether the person could take part in a transfer. If he could, then mark the row he is in, and if not, then mark his column. The decisive point is whether there remains a circled cell outside any of the marked lines. If there is no such circle, then by König's theorem the largest possible number has been assigned.

If there is an unassigned ringed cell, say R, outside any of the marked lines, then this can only be so within one of the portions of configurations in Fig. 6–6. But (b) is impossible, because the marked

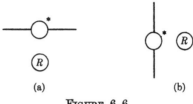

(a)   (b)

FIGURE 6–6

circle could take part in a transfer and would have to be crossed by a horizontal line. Only (a) remains as a possible case and the transfer indicated can be carried out, after which R can be given an additional asterisk.

EXAMPLE 6–9 (same assumptions as in Example 6–8).

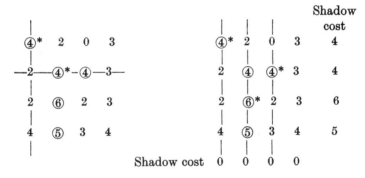

The three lines cover all circles, and we have therefore marked the largest possible number of ringed cells. This brings us to the same intermediate stage which appeared also during the previous procedure. However, we have reached it by a different argument.

The improvement of the solution to the original dual, i.e. the creation of more ringed cells, depends once more on the shadow costs and on additions to and subtractions from them, but while previously they were suggested by the labels, now they are suggested by the marked lines.

If a constant t is added to the shadow costs of all marked columns, and subtracted from all those of unmarked rows, then a marked cell remains ringed, since either its row or its column is marked, but not both. If t equals the smallest difference $u_i + v_j - r_{ij}$ of any cell that is not in a marked row or column, then at least one more cell will be ringed and the new u_i and v_j remain solutions of the dual.

Thus we proceed by further applications of the Primal-Dual algorithm to the final solution.

EXAMPLE 6–9 (continued). The smallest of

$$4 - 3$$
$$4 - 3$$
$$6 - 3$$
$$5 - 4$$

is 1, whence we obtain

				Shadow cost				
④*	2	0	③	3	④*	2	0	③
2	④	④*	③	3	2	④	④*	③
2	⑥*	2	3	5 →	2	⑥*	2	3
4	⑤	3	④	4	4	⑤	3	④*

Shadow cost   1   1   1   0   Optimal.

It follows from the theory of the Primal-Dual algorithm that this procedure terminates after a finite number of steps. E. Egerváry, in a Hungarian paper translated by H. W. Kuhn [60], argued as follows (expressed in terms of our present concepts).

The totals of the shadow costs of crossed rows and columns are reduced every time we apply the second part of the routine. If there are m assignments, and r rows and hence $m - r$ columns are crossed, then we reduce $n - r$ shadow costs by t and increase $m - r$ of them by the same amount. There will be a total reduction of $t(n - m)$ which is positive unless n assignments have already been made. Notice also that the said total is larger than the total assigned rating marks when m is less than n, because only m pairs will be used to make up these rating marks, while there are n altogether, and the remaining ones must add up to a positive number, or they would not be solutions of the dual. If n assignments are made, the total of all shadow costs equals that of the assigned rating marks.

Because at each application of this routine the total of the shadow costs is reduced, and because it cannot be reduced without limit, a stage will be reached when the two totals are equal and the problem is solved.

There exists also a version of the method where $n - 1$ lines are always crossed (cf. [97]).

6–7 The bottleneck assignment problem. In a report of the RAND Corporation, O. Gross introduced the following problem: Let n persons be assigned to n jobs, and let rating marks be given, as in the assignment problem. This time, however, we want to make the assignment in such a way that the smallest rating realized is as high as possible.

He points out that after making some assignment, we can find out whether it is the best, according to the requirement just stated, by noting the bottleneck, i.e. the smallest allocated rating mark and then constructing another table, in which we have replaced the merit rating mark by 1 if it exceeds the bottleneck, and by 0 otherwise.

Solve the assignment problem for the modified table and see whether the maximum total of assigned ratings is n. This is the case if there are n independent 1's in the modified table. If this is not so, then the bottleneck problem has been solved, but if there are n independent 1's in the modified table, then using their cells for job allocations will be an assignment with a larger bottleneck, and should replace the previous assignment scheme. Thus we proceed and it is obvious that we must reach the final answer after a finite number of stages.

EXAMPLE 6–10.

Table of merit ratings
and first allocation   Modified table

4*	2	0	3	1	0	0	1
2	4*	4	3	0	1	1	1
2	6	2*	3	0	1	0	1
4	5	3	4*	1	1	1	1 .

We must now solve the assignment problem for the modified table. In the present case we can do it by inspection. Thus

1*	0	0	1
0	1	1	1*
0	1*	0	1
1	1	1*	1 .

Hence a better allocation is the following, leading to the modified table and a solution for the latter, as indicated:

4*	2	0	3	1*	0	0	0
2	4	4	3*	0	1	1*	0
2	6*	2	3	0	1*	0	0
4	5	3*	4	1	1	0	1*

and finally

$$
\begin{array}{cccc}
4^* & 2 & 0 & 3 \\
2 & 4 & 4^* & 3 \\
2 & 6^* & 2 & 3 \\
4 & 5 & 3 & 4^*.
\end{array}
$$

6–8 Multiple distribution. It is natural to consider also groupings according to three characteristics which lead to constraints

$$\sum_i x_{ijk} = a_{jk}, \qquad \sum_j x_{ijk} = b_{ik}, \qquad \sum_k x_{ijk} = c_{ij} \qquad \text{(A)}$$

or

$$\sum_i \sum_j x_{ijk} = a_k, \qquad \sum_i \sum_k x_{ijk} = b_j, \qquad \sum_j \sum_k x_{ijk} = c_i, \qquad \text{(B)}$$

for non-negative variables x_{ijk}.

For (A) to be feasible it is necessary that

$$\sum_k b_{ik} = \sum_j c_{ij}, \qquad \sum_i c_{ij} = \sum_k a_{jk}, \qquad \sum_j a_{jk} = \sum_i b_{ik},$$

but this is not sufficient, as Fig. 6–7, from [111], shows.

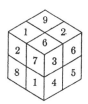

FIGURE 6–7

It follows from the constraints that

$$x_{ijk} \leqslant \min (a_{jk}, b_{ik}, c_{ij}) = m_{ijk}, \text{ say,}$$

for all i, j, k, and hence it is also necessary that

$$a_{jk} \leqslant \sum_i m_{ijk}, \qquad b_{ik} \leqslant \sum_j m_{ijk}, \qquad c_{ij} \leqslant \sum_k m_{ijk}.$$

The last inequality is not satisfied in the above example for $i = j = 1$.

The sets which satisfy either (A) or (B) do not always have only vertices with integer coordinates. For (B) this is exemplified by the Travelling Salesman problem (see Section 10–1).

EXERCISES

6–1. Solve the following problem by the network flow method (labelling, Primal-Dual algorithm):

		Jobs					
		1	2	3	4	5	6
	1	0	35	0	31	0	0
	2	35	0	35	22	30	0
Candidates	3	31	22	0	0	31	26
	4	14	0	0	10	0	0
	5	0	30	16	29	0	30
	6	0	0	16	26	36	0.

6–2. Solve the problem of Exercise 6–1 by the Hungarian method.

6–3. Solve the following assignment problem by the Perturbation method:

$$
\begin{array}{cccc}
4 & 2 & 0 & 3 \\
2 & 4 & 4 & 3 \\
2 & 6 & 2 & 3 \\
4 & 5 & 3 & 4.
\end{array}
$$

6–4. Solve the shortest path problem of Example 4–1 as a transshipment problem.

6–5. The following combined cost and requirement table is given:

	2	3	5	3	2
1	8	9	7	6	8
7	7	5	5	3	7
7	12	8	5	8	5.

Which constants should be added to or subtracted from rows and columns so that only cells with zero cost contain positive entries in the optimal solution?

6–6. Six jobs, each taking two hours, can be done either on machine A or on machine B. The times available on both machines are 9–11 a.m., 1–3 p.m., and 5–7 p.m. All jobs must be done on the same day, by the cheapest distribution over times and machines. The cost of a two-hourly period is given in the following table:

On machine	9–11	1–3	5–7
A	1, 2, 3, 2, 1, 3	1, 3, 2, 3, 3, 1	2, 1, 3, 1, 1, 2
B	2, 3, 1, 3, 2, 1	3, 1, 2, 2, 2, 3	3, 2, 1, 3, 3, 2

where the six entries in each block are the costs of jobs 1, 2, . . ., 6 respectively, in this order.

6–7. The "Caterer Problem" [89]. A caterer must provide the following number of clean table napkins on seven consecutive days:

$$120, \quad 60, \quad 70, \quad 100, \quad 90, \quad 70, \quad 110.$$

Soiled napkins can be sent to a laundry which returns them on the third day after, or on the next day, for a higher cost. The cost of a new napkin is 30 cents, that of the fast service is 15 cents, and that of the slow service 5 cents, per napkin. How many napkins should be bought, and how many sent, each day, to the slower and to the faster service?

6–8. Solve the transportation problem for the following cost and requirements table:

	1	1	1	1	1	1
1	9	22	58	11	19	27
1	43	78	72	50	63	48
1	41	28	91	37	45	33
1	74	42	27	49	39	32
1	36	11	57	22	25	18
1	3	56	53	31	17	28.

6–9. Solve the following transportation problem:

	15	20	30	35
25	10	5	6	7
25	8	2	7	6
50	9	3	4	8.

(This is the first published Transportation Problem. It is the example in [85].)

6–10. Transform the capacitated transportation problem of Example 6–6 into an ordinary transportation problem by Harvey's method, and solve it, avoiding degeneracy.

CHAPTER 7

USES OF DUALITY. ECONOMIC INTERPRETATION

The concept of duality is fundamental to an understanding of algebraic relationships between inequalities. It has also important practical applications.

If it is easier to solve the dual of that system in which we are primarily interested, then we shall solve the dual, and we shall be able to extract all the relevant information from its final tableau. For instance, if we want to use the Simplex method, then it is easier to start from inequalities of such a type that the slack variables can be used to form the first set of basic feasible variables.

It is sometimes useful to remember that if we use the Simplex method in both systems, then the common optimal value of the objective functions is approached from above in the minimizing, and from below in the maximizing problem. It follows that if we are merely interested in knowing whether, for instance, the latter will reach or exceed a certain value, say B_0, then we might obtain this knowledge more quickly from the dual system; if its objective function has become smaller than B_0, then we know that the common optimal value cannot possibly reach B_0.

If we are forced to add a new constraint to a problem which has already been solved, then it would seem that we have to start afresh, because a feasible solution reached is not necessarily feasible in the new problem as well. But taking the dual, we have to add a new variable, and the previous feasible solution remains feasible when we simply set the new variable equal to zero.

7–1 Shadow prices. The most valuable practical use depends, however, on the fact that the final tableau of the Simplex method contains the optimal value of the objective function and that this value is

$$c_1 x_1^0 + \ldots + c_m x_m^0,$$

as well as

$$b_1 z_{0m+1} + \ldots + b_n z_{0N}$$

(assuming that the final basic variables of the primal are $x_1, \ldots, x_m$). This allows us to answer the question of how much it is worth paying for additional facilities in a production process.

EXAMPLE 7–1. Consider the problem of the manufacturer mentioned at the end of Chapter 1. It is the problem of Example 5–1, when the right-hand side has been multiplied by 10. The final tableau is then

		x_4	x_5	x_3	2
3	x_1	0.6	−0.8	−1	4
6	x_2	−0.2	0.6	1	2
	B	0.6	1.2	1	24

We have, indeed,

$$3 \times 4 + 6 \times 2 = 20 \times 0.6 + 10 \times 1.2 = 24.$$

Imagine, now, that we could obtain more material, in addition to $c_1 (= 20)$ and $c_2 (= 10)$ which we have already got. Then the values of x_1 and of x_2, and hence also of B, could be different in the final tableau. However, as long as x_1 and x_2 remain the basic variables, the remainder of the tableau is not altered. Thus, if c_1 were increased by one unit, the profit $c_1 z_{04} + c_2 z_{05}$ would be increased by $z_{04} (= 0.6)$. This is then the economic meaning of that entry: It is the price that we should be prepared to pay for one additional yard of red wool. Similarly, we could pay up to $z_{05} (= 1.2)$ for one more yard of green wool.

If we had 21 yards of red wool at our disposal, so that this would be the right-hand side of the first constraint, then the final tableau would read

		x_4	x_5	x_3	2
3	x_1	0.6	−0.8	−1	4.6
6	x_2	−0.2	0.6	1	1.8
	B	0.6	1.2	1	24.6

The profit has been increased by 0.6.

We repeat that this is only true as long as the set of basic variables does not change. However, for sufficiently small changes in the c_i this will be so. This explains why we call the values $z_{0,m+j}$ *marginal*, or *shadow*, *prices*. They have also been called *opportunity costs*, or *multipliers*.

It is, of course, easy to determine whether the basis changes as a consequence of changes either in the right-hand side or of the coefficients of the objective function. The relations (2–a) and (2–d) in Chapter 2 show whether the x_i^0 or the $z_{0,m+j}$ have the signs which indicate that we are still dealing with a feasible and optimal solution.

Whenever a constraint is satisfied as a strict inequality, i.e. with positive slack variables, then the corresponding shadow price is zero. This makes economic sense: If a facility is not fully exploited in the most economical scheme, then it is not worth paying anything for its increase.

If the price of one unit of the jth facility is y_{n+j}, and if for one unit of the ith commodity to be produced we need a_{ij} of that facility, then this production would cost

$$\sum_j a_{ij} y_{n+j}.$$

The Duality theorem tells that if

$$y_{n+j} = z_{0,n+j},$$

as defined in the optimal tableau, then the total price $\sum_j c_j y_{n+j}$ of all facilities equals the total income $\sum_i b_i x_i$. The highest acceptable values of the y_{n+j} satisfy also the dual equations $\sum_j a_{ij} y_{n+j} \geqslant b_i$.

We know, also, that if any of the latter inequalities is strictly satisfied, then the ith commodity, which would produce less income than it costs, will not be produced at all.

Even if we do not know the final tableau of a problem, but only know which variables are basic, then we can find the optimal values y_{n+j}^0 of the main variables of the dual problem. Let the optimal basic variables be $x_1, \ldots, x_m$. Then from (2–a)

$$x_i^0 = \sum_{t=1}^m d_{ti} b_t \qquad (i = 1, \ldots, m),$$

and hence

$$\sum_{i=1}^m c_i x_i^0 = \sum_{i=1}^m \sum_{t=1}^m c_i d_{ti} b_t = \sum_{t=1}^m y_{n+t}^0 b_t.$$

Therefore

$$y_{n+t}^0 = \sum_{i=1}^m c_i d_{ti}.$$

This *Regrouping Principle* (see [36]) is, of course, particularly useful when the solution of the primal problem is analytically given.

EXAMPLE 7–2. In Example 5–1 we have found that the final basic variables of the problem

$$3x_1 + 4x_2 + x_3 + x_4 = 2$$
$$x_1 + 3x_2 + 2x_3 + x_5 = 1$$
$$\text{Maximize } 3x_1 + 6x_2 + 2x_3$$

are x_1 and x_2. The system

$$3x_1 + 4x_2 = b_1, \qquad x_1 + 3x_2 = b_2$$

is solved by

$$x_1 = \frac{3b_1 - 4b_2}{5}, \qquad x_2 = \frac{-b_1 + 3b_2}{5}.$$

The value of

$$3x_1 + 6x_2 + 2x_3 = \frac{3b_1 + 6b_2}{5}$$

must equal that of

$$b_1 y_4 + b_2 y_5,$$

so that $y_4 = 0.6$ and $y_5 = 1.2$ are the optimal values of the main variables of the dual problem.

7–2 Efficient points. Let a system

$$a_{1j}x_1 + \ldots + a_{nj}x_n \leqslant b_j \qquad (j = 1, \ldots, m)$$
$$x_i \geqslant 0 \qquad\qquad\qquad (i = 1, \ldots, n)$$

be given. A point $(x_1^0, \ldots, x_n^0)$ is called *efficient* if there is no point $(x_1^0 + y_1, \ldots, x_n^0 + y_n)$ with all y_i non-negative, which satisfies the constraints, except for $y_i = 0$ for all i.

Given $x_1^0, \ldots, x_n^0$, we can find out whether this is an efficient point by solving the scheme

$$a_{1j}(x_1^0 + y_1) + \ldots + a_{nj}(x_n^0 + y_n) \leqslant b_j,$$
$$y_i \geqslant 0.$$
$$\text{Maximize } y_1 + \ldots + y_n.$$

If, and only if, the maximum is zero, then the point is efficient.

The problem of finding an efficient point to begin with is not quite so straightforward. It means that we want to find conditions for $x_1^0, \ldots, x_n^0$ which ensure that the maximum of $y_1 + \ldots + y_n$ is zero. An example will make the procedure clear.

EXAMPLE 7–3. Given

$$-3x_1 - x_2 \leqslant -3,$$
$$4x_1 + 3x_2 \leqslant 6,$$
$$- x_1 - 2x_2 \leqslant -2.$$

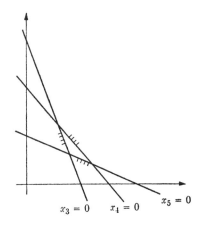

$x_3 = 0$ $x_4 = 0$ $x_5 = 0$

This leads to

	y_1	y_2	
y_3	-3	-1	$-3 + 3x_1^0 + x_2^0$
y_4	4	3	$6 - 4x_1^0 - 3x_2^0$
y_5	-1	-2	$-2 + x_1^0 + 2x_2^0$
	-1	-1	0

Only those points can be efficient whose coordinates make the value of y_4 zero, because there are positive values in that row, and using one of them as a pivot would, otherwise, increase $y_1 + y_2$ from its present value, zero. These are the points on the line $x_4 = 0$. Using the intersection of row y_4 and column y_1 as a pivot, we obtain the next tableaux:

	y_4	y_2	
y_3	$3/4$	$5/4$	$-3 + 3x_1^0 + x_2^0$
y_1	$1/4$	$3/4*$	0
y_5	$1/4$	$-5/4$	$-2 + x_1^0 + 2x_2^0$
	$1/4$	$-1/4$	0

,

	y_4	y_1	
y_3	$1/3$	$-5/3$	$-3 + 3x_1^0 + x_2^0$
y_2	$1/3$	$4/3$	0
y_5	$2/3$	$5/3$	$-2 + x_1^0 + 2x_2^0$
	$1/3$	$1/3$	0

.

Hence all those points on the line $x_4 = 0$ between (and including) the intersections with the lines $x_3 = 0$ and $x_5 = 0$ are efficient. (Confirm from the figure!)

In a different context, Koopmans (in [10], Chapter 3) calls the point $(b_1, \ldots, b_m)$ efficient, if the constraints

$$a_{1j}x_1 + \ldots + a_{mj}x_m = b_j + z_j \quad (j = 1, \ldots, m),$$
$$x_i \geqslant 0 \text{ (all } i) \quad \text{and} \quad z_j \geqslant 0 \text{ (all } j)$$

can only be solved by $z_1 = \ldots = z_m = 0$ or, equivalently, if the minimum of $-v(z_1 + \ldots + z_m)$ is zero, where v is any positive number.

By duality, this is equivalent to demanding that there should be a solution to

$$a_{i1}y_1 + \ldots + a_{im}y_m \leqslant 0 \quad (i = 1, \ldots, n),$$
$$y_j \geqslant v > 0 \, (j = 1, \ldots, m),$$

which makes $b_1y_1 + \ldots + b_my_m = 0$, because the latter expression can only be $\leqslant 0$ (since $-(z_1 + \ldots + z_m) \leqslant 0$ for any set of $z_j \geqslant 0$).

7–3 Input-output analysis. One of the inspirations for the development of linear programming studies was the analysis of interindustry relationships, as introduced by W. W. Leontief in [12]. It is concerned with the interrelationship between activities, due to the competition for scarce resources.

EXAMPLE 7–4. This example is adapted from T. Barna [22]. Assume that for the production of one unit of machinery one needs 0.2 unit of coal, and 0.5 unit of steel, that for the production of one unit of steel one needs 2 units of coal and 0.3 unit of machinery, and that for producing one unit of coal 0.1 unit of steel and 0.1 unit of machinery are required. For the sake of simplicity we ignore here all other factors.

Assume that one unit of machinery should be sold on the market. Then, denoting the numbers of units of coal, steel, and machinery by c, s, and m, respectively, we have the following relations:

$$c - 2.0s - 0.2m = 0,$$
$$s - 0.1c - 0.5m = 0,$$
$$m - 0.1c - 0.3s = 1,$$

and hence

$$c = 2.31, \quad s = 1.00, \quad m = 1.53.$$

In the Leontief equilibrium model of an economy it is assumed that there exists a set of activities (for instance, industries) and that each of these has one single type of output. If one unit effort of the ith activity produces one unit of the jth type of output, using for this purpose outputs of other activities, then we can describe this by a column of constants, with $a_{i,j_i} = 1$ in the j_ith row, and zero or negative values in the other places, indicating the corresponding amounts of items which are consumed in that activity.

Though each activity is supposed to produce just one single type of output, the same type could be produced by more than one activity. We refer then to *a model with substitution.* Given that b_j units are required of the jth type of item, we have to find the levels x_i of activity i such that

$$\sum_{i=1}^n a_{ij}x_i = b_j \qquad (j = 1, \ldots, m),$$

where the set b_j, consisting of positive values, is called the *Bill of Goods.*

In a model without substitution we have $n = m$, in one with substitution $n > m$, and one can then introduce an optimizing condition, e.g. $\sum_i c_i x_i$ to be minimized. Here c_i is, perhaps, the labour requirement for one unit of the activity i. The fact that each variable has a positive coefficient in only one constraint leads to conclusions of economic interest. For instance, it can be shown that the set of optimal basic variables (though not, of course, their values) is independent of the Bill of Goods. We do not pursue this matter further, but refer the reader to the literature (see, for instance, [42]). An extensive development of Linear Programming from the point of view of economists is contained in [6].

EXERCISES

7–1. How does the minimum of $x_1 - x_2$ change when the right-hand side of the second inequality in Exercise 5–1 is changed (a) to 2, (b) to 4?

7–2. Consider the numerical example at the end of Chapter 1. How much should we be prepared to pay for another ten yards of (a) red wool, (b) green wool?

7–3. A unit of product A, B, or C sells at 5, 3, or 4 respectively. The following table shows how much of raw materials a or b is required for one unit of A, B, or C, and also how much of a and b is altogether available.

	A	B	C	Available
a	3	2	3	12
b	4	1	2	15

Impute costs to the raw materials such that the total cost of all material equals the total price obtained for the finished commodities, and such that no commodity contains material of less imputed cost than the selling price of that commodity.

CHAPTER 8

SELECTED APPLICATIONS

In this chapter we present applications of the algorithms we have dealt with to problems of industry and to other branches of applied mathematics. These problems will be worked out in some detail, to give the reader an idea of the scope and range of mathematical techniques of planning.

8–1 The nutrition problem. This problem appeared first in a paper by G. J. Stigler [114]. After formulating the problem in linear programming terms, he said, "The procedure is experimental because there does not appear to be any direct method of finding the minimum of a linear function subject to linear conditions." At present we do have such a direct method, and we shall apply it to an illustrative problem.

The following nutritional values are assumed to be necessary per day:

70 gm of proteins,

3000 calories,

800 mg of calcium,

12 mg of iron.

A shopper finds that the following foods are available on the market:

(1) Brown bread from mixed grist flour
at a price of 3d. per 100 gm
(2) Cheddar cheese at a price of 7d. per 100 gm
(3) Fresh butter at a price of   7d. per 100 gm
(4) Baked beans at a price of   5d. per 100 gm
(5) Spinach at a price of   2d. per 100 gm.

He (or she) wishes to provide the necessary ingredients at the lowest total cost. To decide what to buy, and how much of it, it is clearly necessary to know how much of all ingredients there is in a unit of the food items, and the information of Table 8–1 is extracted from McCance and Widdowson, *Chemical Composition of Foods*, 2nd ed. (M.R.C. 1946). Before we write the constraints which express the necessity of providing the ingredients from the available items, we have to decide

TABLE 8-1

Item	Protein, gm	Calories	Calcium, mg	Iron, mg
(1)	8.3	246	17.2	2.01
(2)	24.9	423	810	0.57
(3)	0.4	793	14.8	0.16
(4)	6	93	61.6	2.05
(5)	5.1	26	595	4

whether we want to provide the ingredients precisely in the necessary minimum amounts, or whether we are prepared to oversupply some of them. The latter does not do any harm (for all we know), and it may be cheaper than to insist on the precise amounts. Imagine, for instance, that some cheap food contains little of ingredient A, but a great amount of B. Then it is quite possible that by providing precisely the minimum amount of A we oversupply B, but that this is still cheaper than any other combination, because of the cheapness of this food item. In fact, it is even conceivable that precise supply of all items is not feasible. To see this, imagine that only one single item is available, which does not contain the ingredients in the precise proportion of the required minima. It is easily seen that even if more than one item is offered, such a case can still exist.

Our constraints state, therefore, that at least the necessary minima will be supplied, and the function to be minimized is the cost of all the purchases. Let the numbers of the units of 100 gm of the items bought be denoted by $x_1, \ldots, x_5$. Since the amounts in Table 8-1, which supplies the coefficients in the constraints, vary greatly, we shall divide some of the inequalities by suitable factors, and obtain the set:

$$8.3x_1 + 24.9x_2 + 0.4x_3 + 6x_4 + 5.1x_5 \geqslant 70,$$
$$2.46x_1 + 4.23x_2 + 7.93x_3 + 0.93x_4 + 0.26x_5 \geqslant 30,$$
$$1.72x_1 + 81x_2 + 1.48x_3 + 6.16x_4 + 59.5x_5 \geqslant 80,$$
$$2.01x_1 + 0.57x_2 + 0.16x_3 + 2.05x_4 + 4x_5 \geqslant 12.$$

Minimize $3x_1 + 7x_2 + 7x_3 + 5x_4 + 2x_5$.

This is, then, our linear programming problem that we have to solve.

When we look at this problem more closely, we might feel that we have ignored some conditions which we should certainly like to impose on our diet. For instance, we should not like our diet to consist entirely of brown bread and spinach, even if the quantities of them were sufficient to keep us healthy—in theory. This is a typical anxiety in the mathematical formulation of any problem: Have we really considered all the relevant facts? However, at this stage we explore the consequences of the conditions as we have formulated them, and as it happens, the resulting diet will—to most of us—appear to be quite acceptable.

First, we introduce the additional variables $x_6, \ldots, x_9$, to convert the inequalities into equations. But this does not give us any clue to a first feasible solution in the Simplex method, because these additional variables appear on the left with negative signs, and the constants on the right are positive. Hence we introduce a set of four artificial variables, which we denote by $x_{101}, \ldots, x_{104}$, to distinguish them easily from those variables with some realistic meaning. The artificial variables are introduced, one in each equation, with positive signs, and produce the following first tableau:

		3	7	7	5	2				
		x_1	x_2	x_3	x_4	x_5	x_6	x_7	x_8	x_9
M x_{101}	70	8.3	24.9	0.4	6	5.1	−1			
M x_{102}	30	2.46	4.23	7.93	0.93	0.26		−1		
M x_{103}	80	1.72	81*	1.48	6.16	59.5			−1	
M x_{104}	12	2.01	0.57	0.16	2.05	4				−1
		−3	−7	−7	−5	−2				
M-row	192	14.49	110.7	9.97	15.14	68.86	−1	−1	−1	−1

Since the value of M is taken to be larger than any with which we compare it during the procedure, the M-row is overriding, and the largest positive value in the M-row is that in the x_2 column. We make therefore x_2 basic (remember that we want to minimize the objective function).

The smallest of the ratios 70/24.9, 30/4.23, 80/81, 12/0.57 is the third, belonging to the row of x_{103}. Hence an exchange of x_2 and x_{103} is indicated. It leads to the tableaux of Table 8–2. The fifth tableau contains the answer.

Table 8–3 lists the items we should buy and the ingredients provided by each item. We have an oversupply of iron ($x_9 = 0.485744$). Beans (x_4) and spinach (x_5) are not included in the diet, and there is no other oversupply ($x_6 = x_7 = x_8 = 0$).

TABLE 8–2

2nd tableau

		3 x_1	M x_{103}	7 x_3	5 x_4	2 x_5	x_6	x_7	x_8	x_9
M	x_{101}	45.407408	−.307407	−.054973	4.106380	−13.190743	−1.000000	.000000	.307415	.000000
M	x_{102}	25.822224	−.052222	7.852709	.603313	−2.847223	.000000	−1.000000	.052224	.000000
7	x_2	.987654	.012346	.018272	.076049	.733568	.000000	.000000	−.012346	.000000
M	x_{104}	11.437037	−.007037	.149585	2.006652	3.581296	.000000	.000000	.007037	−1.000000
		6.913578	.086422	−6.872096	−4.467657	3.141976	.000000	.000000	−.086422	
		82.666669	−1.366666	7.947321	6.721345	−12.456670	−1.000000	−1.000000	.366676	

3rd tableau

		M x_{104}	M x_{103}	7 x_3	5 x_4	2 x_5	x_6	x_7	x_8	x_9
M	x_{101}	.920520	−3.889716	−.280037	−.636814	−3.698929	−27.120969	−1.000000	.000000	.280045	3.889719
M	x_{102}	12.254042	−1.186336	−.043874	7.675252*	−1.772251	−7.095844	.000000	−1.000000	.043876	1.186337
7	x_2	0.866096	−.010629	.012421	.016682	.054721	.696504	.000000	.000000	−.012421	.010629
3	x_1	5.724541	.500527	−.003522	.074871	1.004383	1.792534	.000000	.000000	.003522	−.500527
		23.236297	1.427169	.076381	−6.658613	−1.603804	8.253130	.000000	.000000	−.076381	−1.427178
		13.174562	−6.076052	−1.323911	7.038438	−5.471180	−34.316813	−1.000000	−1.000000	.323921	5.076057

160

TABLE 8-2 (continued)

4th tableau

			M x_104	M x_103	M x_102	5 x_4	2 x_5	x_6	x_7	x_8	x_9
M	x_101	1.937222	−3.988146	−.283677	.082970	−3.845973	−27.709710	−1.000000	−.082970	.283686	3.988148*
7	x_3	1.596565	−.154566	−.005716	.130289	−.230905	−.924510	.000000	−.130289	.005717	.154566
7	x_2	.839463	−.008051	.012516	−.002173	.058573	.711927	.000000	.002173	−.012516	.008051
3	x_1	5.605004	.512100	−.003094	−.009755	1.021671	1.861753	.000000	.009755	.003094	−.512100
		33.867207	.397981	.038318	.867547	−3.141311	2.097178	.000000	−.867547	−.038311	−.397974
		1.937222	−4.988146	−1.283677	−.917030	−3.845973	−27.709710	−1.000000	−.082970	.283686	3.988148

5th tableau

			M x_104	M x_103	M x_102	5 x_4	2 x_5	x_6	x_7	x_8	M x_101
	x_9	.485744	−1.000000	−.071130	.020803	−.964355	−6.948014	−.250743	−.020803	.071130	.250743
7	x_3	1.521489	.000000	.005278	.127073	−.081847	.149422	.038756	−.127073	−.005278	−.038756
7	x_2	.835553	.000000	.013087	−.002341	.066337	.767858	.002019	.002341	−.013087	−.002019
3	x_1	5.853754	.000000	−.039519	.000898	.527825	−1.696321	−.128406	−.000898	.039519	.128406
		34.060524	.000000	.010007	.875823	−3.525102	−.668003	−.099792	−.875823	−.010007	.099792
		.000000	−1.000000	−1.000000	−1.000000	.000000	.000000	.000000	.000000	.000000	−1.000000

TABLE 8–3

Item		Weight, gm	Protein, gm	Calories	Calcium, mg	Iron, mg
x_1	Bread	585.3754	48.586	1440.022	100.685	11.766
x_2	Cheese	83.5553	20.805	353.438	676.797	0.477
x_3	Butter	152.1489	0.609	1206.540	22.518	0.243
			70.000	3000.000	800.000	12.486

The dual formulation does not require artificial variables and might therefore appear to be preferable. As a matter of fact, it leads to a higher number of iterations. We would start off with

$$8.3y_6 + 2.46y_7 + 1.72y_8 + 2.01y_9 \leqslant 3,$$
$$24.9y_6 + 4.23y_7 + 81.0y_8 + 0.57y_9 \leqslant 7,$$
$$0.4y_6 + 7.93y_7 + 1.48y_8 + 0.16y_9 \leqslant 7,$$
$$6y_6 + 0.93y_7 + 6.16y_8 + 2.05y_9 \leqslant 5,$$
$$5.1y_6 + 0.26y_7 + 59.5y_8 + 4y_9 \leqslant 2.$$

Maximize $70y_6 + 30y_7 + 80y_8 + 12y_9 = B$, say.

We quote the basic variables and their values as they would appear in the successive tableaux; together with the objective function

					B	
$y_1 = 3$	$y_2 = 7$	$y_3 = 7$	$y_4 = 5$	$y_5 = 2$	0	(1)
2.942	4.278	6.950	4.793	$y_8 = 0.034$	2.689	(2)
1.000	$y_6 = 0.238$	6.885	3.490	0.013	17.729	(3)
0.401	0.278	6.866	3.034	$y_9 = 0.146$	21.195	(4)
$y_7 = 0.301$	0.226	4.494	2.971	0.193	27.129	(5)
0.874	0.076	$y_8 = 0.017$	3.432	0.096	34.014	(6)
0.876	0.100	0.010	3.525	$y_5 = 0.668$	34.061.	(7)

These are, of course, also the absolute values of the corresponding shadow costs in the primal tableaux.

8–2 A paper-trim problem. A machine for making paper produces reels of a width of 82 in. and the following orders have been received:

$$60 \pm 6 \text{ lb of width 58 in.,}$$
$$85 \pm 8\tfrac{1}{2} \text{ lb of width 26 in.,}$$
$$85 \pm 17 \text{ lb of width 24 in.,}$$
$$50 \pm 10 \text{ lb of width 23 in.}$$

It is required to satisfy these orders in such a way that the relative trim waste from reels with width of 82 in. is as small as possible.

To obtain a survey of the possibilities we list all combinations of the required widths contained in 82 in., ignoring combinations for which the trim waste would exceed 23 in., the smallest width on order (Table 8–4).

TABLE 8–4

Combination	58 in.	26 in.	24 in.	23 in.	Wastage, in.
1	1		1		0
2	1			1	1
3		3			4
4		2	1		6
5		2		1	7
6		1	2		8
7		1	1	1	9
8		1		2	10
9			3		10
10			2	1	11
11			1	2	12
12				3	13

In order to deal with consistent units we take as the unit of a reel that length, or diameter, which gives to a reel of 82 in. a weight of 82 lb. Then

$$60 \text{ lb of width 58 in. correspond to } 1.03448 \text{ reels.}$$
$$85 \text{ lb of width 26 in. correspond to } 3.26923 \text{ reels.}$$
$$85 \text{ lb of width 24 in. correspond to } 3.54167 \text{ reels.}$$
$$50 \text{ lb of width 23 in. correspond to } 2.17391 \text{ reels.}$$

On these amounts tolerances of 10% and 20% are allowed, which produce the upper and lower limits of Table 8–5. We denote the

TABLE 8–5

	58 in.	26 in.	24 in.	23 in.
j	1	2	3	4
L_j	0.93103	2.94231	2.83334	1.73913
U_j	1.13793	3.59615	4.25000	2.60869

number of reels cut according to combination i by x_i. Then the amount ordered of width 58 in. can be obtained from combinations 1 and 2, and therefore $x_1 + x_2$ must be at least 0.93103, and everything above 1.13793 is a waste.

We introduce y_j as slack variables and have then the following constraints:

$$x_i \geqslant 0, \qquad y_j \geqslant 0$$

$$x_1 + x_2 - y_1 \qquad\qquad\qquad\qquad\qquad = 0.93103, \quad \text{(a')}$$

$$3x_3 + 2x_4 + 2x_5 + x_6 + x_7 + x_8 - y_2 \qquad = 2.94231, \quad \text{(b')}$$

$$x_1 \quad + x_4 \quad + 2x_6 + x_7 \quad + 3x_9 + 2x_{10} + x_{11} - y_3$$
$$= 2.83334, \quad \text{(c')}$$

$$x_2 \quad + x_5 \quad + x_7 + 2x_8 \quad + x_{10} + 2x_{11} + 3x_{12} - y_4$$
$$= 1.73913. \quad \text{(d')}$$

We want to minimize the wastage; this is composed of the trim waste listed in Table 8–4 according to combinations, and also of those reels cut which are in excess (if any) of the tolerances of the orders. The latter equal $y_j - (U_j - L_j)$ for each pound of reel, provided this expression is positive. Otherwise there is no waste on this account.

To cope with the nonlinearity implied in the latter condition, we introduce more non-negative variables. Thus

$$y_j - (U_j - L_j) = x_{12+j} - x_{16+j} \qquad (j = 1, 2, 3, 4).$$

Because the y_j are non-negative, the new variables are subject to

$$-x_{12+j} + x_{16+j} \leqslant U_j - L_j,$$

i.e.
$$-x_{13} + x_{17} \leqslant 0.20690, \qquad \text{(e)}$$
$$-x_{14} + x_{18} \leqslant 0.65384, \qquad \text{(f)}$$
$$-x_{15} + x_{19} \leqslant 1.41666, \qquad \text{(g)}$$
$$-x_{16} + x_{20} \leqslant 0.86956. \qquad \text{(h)}$$

The objective function can then be written

$$\begin{aligned}
C' = \ & x_2 + 4x_3 + 6x_4 + 7x_5 + 8x_6 + 9x_7 \\
& + 10x_8 + 10x_9 + 11x_{10} + 12x_{11} + 13x_{12} \\
& + 58x_{13} + 26x_{14} + 24x_{15} + 23x_{16}. \qquad \text{(i)}
\end{aligned}$$

The terms in the first and second rows are due to trim waste, and those in the third row to redundant reels. Since the x_{12+j} have positive coefficients, and since we want to minimize the objective function, these variables will either be zero (so that $y_j - (U_j - L_j) = -x_{16+j}$ must not affect the objective function), or positive with the smallest possible amount, so that $y_j - (U_j - L_j) = x_{12+j}$ and we have again the correct contribution to the objective function.

The right-hand sides of (a') to (d') are the L_j, and if we substitute for the y_j we obtain

$$x_1 + x_2 - x_{13} + x_{17} \qquad\qquad\qquad\qquad = 1.13793, \text{ (a)}$$
$$3x_3 + 2x_4 + 2x_5 + x_6 + x_7 + x_8 - x_{14} + x_{18} \qquad = 3.59615, \text{ (b)}$$
$$x_1 + x_4 + 2x_6 + x_7 + 3x_9 + 2x_{10} + x_{11} - x_{15} + x_{19} = 4.25000, \text{ (c)}$$
$$x_2 + x_5 + x_7 + 2x_8 + x_{10} + 2x_{11} + 3x_{12} - x_{16} + x_{20} = 2.60869. \text{ (d)}$$

If we left it at that, we would always provide the lowest tolerated limit of all orders, because this would minimize the absolute value of the total wastage. But to minimize the relative wastage, we require also that the total weight of paper supplied should be

$$60 + 85 + 85 + 50 = 280 \text{ lb.}$$

This means that all the paper cut, less the wastage, should be 280, i.e.

$$82 \sum_{i=1}^{12} x_i = 280 + C'. \qquad \text{(j)}$$

We might therefore, equivalently, require that

$$x_1 + \ldots + x_{12} = C$$

should be minimized, and the wastage is then $82C - 280$.

The problem is then to minimize C, subject to (a) to (d), (e) to (h), and (j).

TABLE 8-6

	x_2	x_4	x_5	x_6	x_7	x_8	x_{10}	x_{11}	x_{13}	x_{14}	x_{15}	x_{16}	x_{17}	x_{18}	
x_1	1	0	0	0	0	0	0	0	-1	0	0	0	1	0	1.137930
x_3	0	2/3	2/3	1/3	1/3	1/3	0	0	0	$-1/3$	0	0	0	1/3	1.198718
x_9	$-1/3$	1/3	0	2/3	1/3	0	2/3	1/3	1/3	0	$-1/3$	$-23/72$	$-41/36$	$-13/36$	0.738749
x_{19}	0	0	0	0	0	0	0	0	0	0	0	23/24	29/12	13/12	0.895828
x_{12}	1/3	0	1/3	0	1/3	2/3	1/3	2/3	0	0	0	0	0	0	0.579712
z_1	0	0	0	0	0	0	0	0	-1	0	0	0	1	0	0.206900
x_{20}	0	0	0	0	0	0	0	0	0	-1	0	-1	0	0	0.869560
z_2	0	0	0	0	0	0	0	0	0	0	0	0	0	1	0.653840
z_3	0	0	0	0	0	0	0	0	0	0	-1	$-23/24$	$-29/12$	$-13/12$	0.520832
	0	0	0	0	0	0	0	0	$-2/3$	$-1/3$	$-1/3$	$-23/72$	$-5/36$	$-1/36$	3.655109

We reproduce in Table 8–6 one of the final tableaux, ignoring the additional variable z_4. The value of C is 3.655109, and hence the minimal waste is $C' = 19.718$. The schedule of Table 8–7 results from the

TABLE 8–7

	Wastage	58 in.	26 in.	24 in.	23 in.
x_1 1.138	0	66		27.31	
x_3 1.199	4.795		93.5		
x_9 0.739	7.387			53.19	
x_{12} 0.580	7.536				40
	19.718	66	93.5	80.5	40

tableau of Table 8–6. Of width 58 in. 66 lb are on the upper tolerance limit; hence

$$z_1 = U_1 - L_1 \quad \text{and} \quad x_{17} = 0.$$

Of width 26 in. 93.5 lb are also on the upper tolerance limit; hence

$$z_2 = U_2 - L_2 \quad \text{and} \quad x_{18} = 0.$$

Of width 24 in. 80.5 lb are between the two tolerance limits; hence

$$z_3 \neq U_3 - L_3 \quad \text{and} \quad x_{19} \neq 0,$$

but

$$z_3 + x_{19} = U_3 - L_3.$$

Of width 23 in. 40 lb are on the lower tolerance limit; hence

$$z_4 = 0 \quad \text{and} \quad x_{20} = U_4 - L_4.$$

It will be seen that there are many alternative optimal tableaux. We leave their determination to the reader. A paper dealing with the trim problem (without tolerances) is [61].

8–3 Machine sequencing. It is required to process n items on 3 machines, and the processing must be done in sequence, first on the first machine, then on the second, and then on the third. The processing times on the machines are different for different items. They are denoted by t_i^k for the ith item on the kth machine (cf. H. M. Wagner [124]).

The order in which the items are put into the machines determines the total time from start to finishing of all items, and it is this time we want to minimize. It has been shown (S. Johnson [91]) that if there are two or three machines, then nothing can be gained by changing the order of items from one machine to the next, and we assume therefore that the order remains the same throughout.

We introduce the variables x_{ij}, where the first subscript identifies the item, and the second is the ordinal number of the position on any machine. We make $x_{ij} = 1$ if the ith item is put on in the jth position, and $x_{ij} = 0$ otherwise. If we put all the x_{ij} which equal unity in the order of their second subscript, from 1 to n, then the first subscript will be some permutation of these numbers.

Because each item must be processed in some position, we have

$$\sum_j x_{ij} = 1 \qquad \text{for all } i.$$

Because there is some item in each position, we have

$$\sum_i x_{ij} = 1 \qquad \text{for all } j.$$

We must also write the conditions which state (i) that no processing on the second or the third machine can start on any item before that item has been completely processed on the previous machine, and (ii) that only one single item can be processed on the same machine at any given time. For this purpose we introduce three more sets of unknowns (see Figs. 8–1 and 8–2):

b_j^k is the time at which the jth job starts on the kth machine.

u_j^k is the interval between finishing the jth job on the kth machine and starting it on the $(k+1)$th.

r_j^k is the interval between finishing the jth job on the kth machine and starting the $(j+1)$th job on that machine.

It follows from the meaning of x_{ij} that the time required to process the jth ordered job on the kth machine is

$$t_1^k x_{1j} + \ldots + t_n^k x_{nj}.$$

Conditions (i) and (ii) can be formulated as follows:

$$b_j^k = b_j^{k-1} + u_j^{k-1} + \sum_{i=1}^n t_i^{k-1} x_{ij}, \tag{i}$$

$$b_j^k = b_{j-1}^k + r_{j-1}^k + \sum_{i=1}^n t_i^k x_{ij-1}, \tag{ii}$$

for $k = 2, 3$ and $j = 2, \ldots, n$.

EXAMPLE 8–1. Let $n = 2$ and let the values of t_i^k be

$$
\begin{array}{c}
k = \\
\begin{array}{c|ccc}
 & 1 & 2 & 3 \\
\hline
1 & 2 & 1 & 4 \\
2 & 3 & 5 & 2.
\end{array}
\end{array}
$$
$i =$

We have then

$$b_2^2 = b_2^1 + u_2^1 + t_1^1 x_{12} + t_2^1 x_{22},$$
$$b_2^2 = b_1^2 + r_1^2 + t_1^1 x_{11} + t_2^2 x_{21},$$

and

$$b_2^3 = b_2^2 + u_2^2 + t_1^2 x_{12} + t_2^2 x_{22},$$
$$b_2^3 = b_1^3 + r_1^3 + t_1^3 x_{11} + t_2^3 x_{21}.$$

It follows from the definitions that $b_1^2 = b_2^1$, because it would be wasteful not to start the first job on the second machine immediately after it has been finished on the first, and not to start the second job on the first machine when the first has been finished on it. Moreover $b_2^2 - b_1^3 = r_1^2$ (see Fig. 8–2).

Because all coefficients of the unknowns on the right-hand side of the four equations above are positive, we can equate the two right-hand sides for b_2^2, and also those for b_2^3; neither b_2^2 nor b_2^3 is in any danger of becoming negative.

Substituting the numerical values for the t_i^k, we obtain

$$u_2^1 = 3 - 5x_{11} + r_1^2,$$
$$u_2^2 = 1 - 2x_{11} + r_1^3 - r_1^2,$$

and, in addition,

$$x_{11} + x_{12} = x_{11} + x_{21} = x_{21} + x_{22} = 1.$$

We wish to minimize the time taken to the end of the second job on the third machine and this is easily seen to be equal to (see Fig. 8–2)

$$t_1^1 x_{11} + t_2^1 x_{21} + t_1^2 x_{11} + t_2^2 x_{21} + r_1^3 + t_1^3 + t_2^3.$$

In the present case, this is

$$14 + r_1^3 - 5x_{11}.$$

We have five equations for the eight unknowns

$$x_{11},\ x_{12},\ x_{21},\ x_{22},\ r_1^2,\ r_1^3,\ u_2^1,\ \text{and}\ u_2^2.$$

To start with a feasible solution, we decide on some order of the jobs. Let us take item No. 2 first, and item No. 1 last. Then we have the

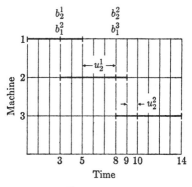

FIGURE 8–1

schedule of Fig. 8–1. For basic variables we choose x_{12}, x_{21}, x_{22}, u_2^1, u_2^2, and have the first tableau

	x_{11}	r_1^2	r_1^3	
x_{12}	1		1	
x_{21}	1		1	
x_{22}	-1		0	
u_2^1	5	-1	3	
u_2^2	2^*	1	-1	1
C	5	0	-1	14

To continue

	u_2^2	r_1^2	r_1^3	
x_{12}	-0.5	-0.5	0.5	0.5
x_{21}	-0.5	-0.5	0.5	0.5
x_{22}	0.5	0.5	-0.5	0.5
u_2^1	-2.5	-3.5	2.5^*	0.5
x_{11}	0.5	0.5	-0.5	0.5
C	-2.5	-2.5	1.5	11.5

	u_2^2	r_1^2	u_2^1	
x_{12}	0	0.2	−0.2	0.4
x_{21}	0	0.2	−0.2	0.4
x_{22}	0	−0.2	0.2	0.6
r_1^3	−1	−1.4	0.4	0.2
x_{11}	0	−0.2	0.2	0.6
C	−1	−0.4	−0.6	11.2

This is the minimum for the continuous case. But we want the answer in integers, of course. Using the discrete programming routine,* we take our cue from the variable x_{12} and add the row

s_1	0	−0.8*	−0.2	−0.6

to obtain

	u_2^2	s_1	u_2^1	
x_{12}	0	1/4	−1/4	1/4
x_{21}	0	1/4	−1/4	1/4
x_{22}	0	−1/4	1/4	3/4
r_1^3	−1	−7/4	3/4	5/4
x_{11}	0	−1/4	1/4	3/4
r_1^2	0	−5/4	1/4	3/4
C	−1	−1/2	−1/2	23/2

Adding the row

s_2	0	−3/4	−1/4	−3/4

we have, finally,

	u_2^2	s_2	u_2^1	
x_{12}	0	1/3	−1/3	0
x_{21}	0	1/3	−1/3	0
x_{22}	0	−1/3	1/3	1
r_1^3	−1	−7/3	4/3	3
x_{11}	0	−1/3	1/3	1
r_1^2	0	−5/3	2/3	2
s_1	0	−4/3	1/3	1
C	−1	−2/3	−1/3	12

---

* Explained in Chapter 10.

This means the schedule of Fig. 8–2 which is shorter than that of Fig. 8–1.

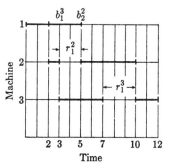

FIGURE 8–2

Another method dealing with this problem is described in [48].

8–4 Approximation to a function. Let the values of a function be $f(x_i)$ at $x = x_i$ ($i = 1, \ldots, m$) and let n functions $g_j(x)$ be given. We should like to approximate the m values $f(x_i)$ by a linear combination of the $g_j(x)$, i.e. by an expression

$$G(x) = t_0 g_0(x) + \ldots + t_n g_n(x),$$

choosing $t_0, \ldots, t_n$ in such a way that the largest deviation between $G(x_i)$ and $f(x_i)$ is as small as possible. Formally, we wish to find the t_j such that the maximum of $|G(x_i) - f(x_i)|$ is minimized.

Let the maximum be denoted by M. Then we have for all i

$$G(x_i) - f(x_i) \leqslant M, \qquad f(x_i) - G(x_i) \leqslant M,$$

and M is to be minimized. Because $G(x_i)$ is a linear expression in the variables t_i, this is again a linear programming problem. Note, however, that it is not required that the variables t_j and M be non-negative, though M will be (cf. [92]).

EXAMPLE 8–2. Let

$$g_j(x) = x^j \qquad \text{for } j = 0, 1, \ldots,$$

and $f(x_i)$ is given in the following table:

x_i	-1	0	1	2
$f(x_i)$	-8	-4	0	$10.$

If we use only $j = 0$, then our problem is: minimize M, subject to

$$M + t_0 \geqslant -8, \qquad M - t_0 \geqslant \quad 8,$$
$$M + t_0 \geqslant -4, \qquad M - t_0 \geqslant \quad 4,$$
$$M + t_0 \geqslant \quad 0, \qquad M - t_0 \geqslant \quad 0,$$
$$M + t_0 \geqslant 10, \qquad M - t_0 \geqslant -10,$$

i.e.

$$M + t_0 \geqslant 10, \qquad M - t_0 \geqslant \quad 8.$$

M is obviously minimized when $t_0 = 1$, $M = 9$ (the largest deviation, see Fig. 8-3).

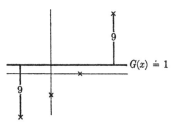

$$G(x) \doteq 1$$

FIGURE 8-3

We shall certainly not go beyond $j = 3$, because a polynomial of order 3 can be found such that all four points lie on its curve. But if we go as far as $j = 3$, then the constraints are as follows:

$$M + t_0 - t_1 + t_2 - t_3 \geqslant -8, \qquad M - t_0 + t_1 - t_2 + t_3 \geqslant \quad 8,$$
$$M + t_0 \qquad\qquad\qquad \geqslant -4, \qquad M - t_0 \qquad\qquad\qquad \geqslant \quad 4,$$
$$M + t_0 + t_1 + t_2 + t_3 \geqslant \quad 0, \qquad M - t_0 - t_1 - t_2 - t_3 \geqslant \quad 0,$$
$$M + t_0 + 2t_1 + 4t_2 + 8t_3 \geqslant 10, \qquad M - t_0 - 2t_1 - 4t_2 - 8t_3 \geqslant -10.$$

We have already dealt with $j = 0$. We continue, to deal with $j = 0$ and 1. Because the Simplex method deals with non-negative variables, we construct the dual to the problem, which contains equations in such variables, viz. (ignoring, for the time being, t_2 and t_3)

Maximize $-8(u_1 - v_1) - 4(u_2 - v_2) + 10(u_4 - v_4) = B$, say,

subject to

$$u_1 + u_2 + u_3 + \quad u_4 + v_1 + v_2 + v_3 + \quad v_4 = 1, \qquad (1)$$
$$u_1 + u_2 + u_3 + \quad u_4 - v_1 - v_2 - v_3 - \quad v_4 = 0, \qquad (2)$$
$$-u_1 + \qquad u_3 + 2u_4 + v_1 - \qquad v_3 - 2v_4 = 0, \qquad (3)$$

It is convenient to consider first the dual to the earlier problem, where only $j = 0$ was used. The objective function remains the same, but only equations (1) and (2) appear as constraints. This is solved by

		-8	-4		4		-10	
		u_1	u_2	u_3	v_2	v_3	v_4	
10	u_4	1	1	1				$\frac{1}{2}$
8	v_1				1	1	1	$\frac{1}{2}$
	B	18	14	10	4	8	18	9

To add the constraint (3), we introduce an artificial variable, equal to the right-hand side of (3)

$$w = -u_1 + u_3 + 2u_4 + v_1 - v_3 - 2v_4,$$

i.e. (expressing w in terms of the nonbasic variables only)

$$w = 3/2 - 3u_1 - 2u_2 - u_3 - v_2 - 2v_3 - 3v_4,$$

so that we have

		-8	-4		4		-10	
		u_1	u_2	u_3	v_2	v_3	v_3	
10	u_4	1	1	1				$\frac{1}{2}$
8	v_1				1	1	1	$\frac{1}{2}$
$-M$	w	3	2	1	1	2	3*	3/2
	B	18	14	10	4	8	18	9
	(M)	-3	-2	-1	-1	-2	-3	$-3/2$

		-8	-4		4		
		u_1	u_2	u_3	v_2	v_3	
10	u_4	1	1	1			$\frac{1}{2}$
8	v_1	-1	$-2/3$	$-1/3$	$2/3$	$1/3*$	0
-10	v_4	1	2/3	1/3	1/3	2/3	$\frac{1}{2}$
	B	0	2	4	-2	-4	0

		-8	-4		4	8	
		u_1	u_2	u_3	v_2	v_1	
10	u_4	1	1	1			$\tfrac{1}{2}$
	v_3	-3	-2	-1	2	3	0
-10	v_4	3^*	2	1	-1	-2	$\tfrac{1}{2}$
	B	-12	-6	0	6	12	0

		-10	-4		4	8	
		v_4	u_2	u_3	v_2	v_1	
10	u_4	$-1/3$	$1/3$	$2/3$	$1/3$	$2/3$	$1/3$
	v_3	1	0	0	1	1	$\tfrac{1}{2}$
-8	u_1	$1/3$	$2/3$	$1/3$	$-1/3$	$-2/3$	$1/6$
	B	4	2	4	2	4	2

To find the corresponding solution to the primal, we consider those constraints which correspond to the basic variables as equations:

$$M + t_0 + 2t_1 = 10, \qquad M - t_0 - t_1 = 0, \qquad M + t_0 - t_1 = -8,$$

so that

$$M = 2 \text{ (as } B \text{ above)}, \qquad t_0 = -4, \qquad t_1 = 6.$$

We draw the curve $y = -4 + 6x$ (Fig. 8–4).

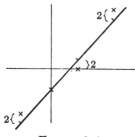

FIGURE 8–4

Adding one more variable in the primal problem, viz. t_2, means adding one more equation to the dual; thus

$$u_1 + u_3 + 4u_4 - v_1 - v_3 - 4v_4 = 0. \tag{4}$$

We introduce again an artificial variable equal to the left-hand side, and express it in terms of the nonbasic variables of the last tableau.

Thus
$$w = 1 - 2u_2 - 2u_3 - 2v_1 - 2v_4.$$
Hence

		-10	-4		4	8	
		v_4	u_2	u_3	v_2	v_1	
10	u_4	$-1/3$	$1/3$	$2/3$	$1/3$	$2/3$	$1/3$
	v_3	1	0	0	1	1	$\frac{1}{2}$
-8	u_1	$1/3$	$2/3$	$1/3$	$-1/3$	$-2/3$	$1/6$
$-M$	w	2	2	2^*	0	2	1
B		4	2	4	2	4	2
(M)		-2	-2	-2	0	-2	-1

		-10	-4		4	8	
		v_4	u_2		v_2	v_1	
10	u_4	-1	$-1/3$		$1/3$	0	0
	v_3	1	0		1	1	$\frac{1}{2}$
-8	u_1	0	$1/3^*$		$-1/3$	-1	0
	u_3	1	1		0	1	$\frac{1}{2}$
B		0	-2		2	0	0

		-10	-8		4	8	
		v_4	u_1		v_2	v_1	
10	u_4	-1	1		0	-1	0
	v_3	1	0		1	1	$\frac{1}{2}$
-4	u_2	0	3		-1	-3	0
	u_3	1	-3		1	4^*	$\frac{1}{2}$
B		0	6		0	-6	0

		-10	-8		4		
		v_4	u_1		v_2	u_3	
10	u_4	$-3/4$	$1/4$		$1/4$	$1/4$	$1/8$
	v_3	$3/4$	$3/4$		$3/4$	$-1/4$	$3/8$
-4	u_2	$3/4$	$3/4$		$-1/4$	$3/4$	$3/8$
8	v_1	$1/4$	$-3/4$		$1/4$	$1/4$	$1/8$
B		$3/2$	$3/2$		$3/2$	$3/2$	$3/4$

To solve the primal, write

$$M + t_0 = -4, \qquad\qquad M - t_0 + t_1 - t_2 = 8,$$
$$M + t_0 + 2t_1 + 4t_2 = 10, \qquad M - t_0 - t_1 - t_2 = 0,$$

so that

$$M = \frac{3}{4}, \qquad t_0 = \frac{-19}{4}, \qquad t_1 = 4, \qquad t_2 = \frac{3}{2},$$

$$y = -\frac{19}{4} + 4x + \frac{3x^2}{2} \quad \text{(Fig. 8-5)}.$$

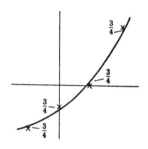

FIGURE 8-5

Going yet one step further, we have

$$-u_1 + u_3 + 8u_4 + v_1 - v_3 - 8v_4 = 0. \tag{5}$$

The new artificial variable is

$$w = \tfrac{3}{4} - \tfrac{3}{2}(u_1 + u_3 + v_2 + v_4),$$

and

		-10	-8	4		
		v_4	u_1	v_2	u_3	
10	u_4	$-3/4$	$1/4$	$1/4$	$1/4$	$1/8$
	v_3	$3/4$	$3/4$	$3/4$	$-1/4$	$3/8$
-4	u_2	$3/4$	$3/4$	$-1/4$	$3/4$	$3/8$
8	v_1	$1/4$	$-3/4$	$1/4$	$1/4$	$1/8$
	w	$3/2*$	$3/2$	$3/2$	$3/2$	$3/4$
B		$3/2$	$3/2$	$3/2$	$3/2$	$3/4$
(M)		$-3/2$	$-3/2$	$-3/2$	$-3/2$	$-3/4$

		u_1	v_2	u_3	
		-8	4		
10	u_4	1	1	1	$\frac{1}{2}$
	v_3	0	0	-1	0
-4	u_2	0	-1	0	0
8	v_1	-1	0	0	0
-10	v_4	1	1	1	$\frac{1}{2}$
B		0	0	0	0

The maximum deviation is zero, i.e. all constraints of the primal are satisfied as equations, which gives

$$t_0 = -4, \qquad t_1 = 3, \qquad t_2 = 0, \qquad t_3 = 1.$$

The curve $y = -4 + 3x + x^3$ passes through all four given points (Fig. 8–6).

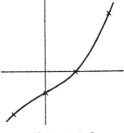

FIGURE 8–6

As a matter of interest, in Table 8–8 we compare the results with those found by least-squares regression, where the sum of squares of the deviations is minimized (compare Exercise 12–1).

TABLE 8–8

j	Minimax solution			Regression		
	Largest deviation	Function	Sum of squares	Largest deviation	Function	Sum of squares
0	9	1	188	10.5	$-\frac{1}{2}$	179
0, 1	2	$-4 + 6x$	12	2.4	$-3.4 + 5.8x$	10.8
0, 1, 2	0.75	$-4.75 + 4x + 1.5x^2$	2.25	0.9	$-4.9 + 4.3x + 1.5x^2$	0.9
0, 1, 2, 3	0	$-4 + 3x + x^3$	0	0	$-4 + 3x + x^3$	0

8–5 Production patterns. A manufacturer knows that in the next n months the requirement for his goods will be $r_1, \ldots, r_n$ units respectively. He has to decide how many units he should produce in the first, second, $\ldots$, nth month, taking into account that storage of goods produced in month i for month j costs $(j - i)$ per unit, and that an increase of production from one month to the next costs t per unit. We ignore any possible complications arising from varying production costs (but see Example 8–3, Section 8–6).

Denote the amount produced in the ith month by x_i $(i = 1, \ldots, n)$. If there is no increase in the requirements from one month to the next, then it is clear that the cheapest production scheme is that according to which precisely as much is produced in each month as is then required. Thus no storage cost will be incurred at all, and no cost for increasing production.

Next, we consider the case when the requirements remain constant or increase from one month to the next. We shall be able to express the result analytically, as distinct from merely arriving at a numerical answer.

We introduce the following notation:

$$\sum_{t=1}^{i} (x_t - r_t) = s_i$$

(the amount stored from the ith to the $(i + 1)$th month).

To cope with the question of increasing production, we write

$$x_i - x_{i-1} = y_i - z_i \ (i = 1, \ldots, n), \quad x_0 = z_1 = 0, \text{ all } y_i, z_i \geqslant 0.$$

We want to minimize

$$\sum_{t=1}^{n-1} s_i + t \sum_{i=1}^{n} y_i = C, \text{ say.}$$

If $x_i - x_{i-1}$ is positive, then $z_i = 0$ and $y_i = x_i - x_{i-1}$ in the optimal solution, while if $x_i - x_{i-1}$ is negative, then $y_i = 0$ and there will be no contribution to the objective function.

The unknowns are the s_i, y_i, and z_i, while the r_i are known constants. The simplest first feasible solution consists of making all y_i basic, and we obtain after simple rearrangements

$$y_1 = r_1 + s_1 + z_1,$$
$$y_2 = -r_1 + r_2 - 2s_1 + s_2 + z_2,$$
$$y_3 = -r_2 + r_3 + s_1 - 2s_2 + s_3 + z_3,$$
$$\vdots$$
$$y_n = -r_{n-1} + r_n + s_{n-2} - 2s_{n-1} + s_n + z_n,$$
$$C = s_1 + \ldots + s_{n-1} + t(r_n + s_n - s_{n-1} + z_1 + \ldots + z_n).$$

If $0 \leqslant t \leqslant 1$, then all coefficients in this expression for C are non-negative, and the optimum has been obtained, since by our assumptions all y_i have non-negative values. However, if t exceeds 1, then this solution is not optimal, and we prove the following theorem: If the r_i are not decreasing, then the optimal solution has for any value of t exceeding 1 the basic variables

$$y_1, \ldots, y_m, s_m, s_{m+1}, \ldots, s_{n-1},$$

where m depends on the value of t.

Solving for these variables, we have $y_1, \ldots, y_{m-1}$ as above, and then

$$(n - m + 1)y_m = \sum_{i=m}^{n} r_i - r_{m-1} + z_m - \sum_{i=1}^{n-m} (y_{m+i} - z_{m+i})(n - m + i + 1)$$
$$- (n - m + 2)s_{m-1} + (n - m + 1)s_{m-2} + s_n,$$

$$(n - m + 1)s_{m+j} = \sum_{i=1}^{n-m-j} r_{m+j+i}(j + 1) - \sum_{i=0}^{j} r_{m+j}(n - m - j)$$
$$- (j + 1) \sum_{i=1}^{n-m-j} (n - m - j - i + 1)$$
$$\times (y_{m+j+i} - z_{m+j+i}) - (n - m - j)$$
$$\times \sum_{i=0}^{j} i(y_{m+i} - z_{m+i}) + (n - m - j)s_{m-1} + (j + 1)s_n.$$

By virtue of our assumption about the r_i this is a feasible solution, and the objective function C consists, apart from terms with obviously positive coefficients, of

$$\sum_{k=1}^{n-m} y_{m+k} \left[\frac{tk}{n - m + 1} - \frac{k(n - m - k + 1)}{2} \right]$$
$$+ s_{m-1} \left[\frac{n - m + 2}{2} - \frac{t}{n - m + 1} \right].$$

Hence all coefficients are non-negative, if

$$\binom{n - m + 1}{2} \leqslant t \leqslant \binom{n - m + 2}{2}.$$

A slight modification is required when $m = 1$, since s_0 is not defined. In this case the basic solution is optimal if $t \geqslant \binom{n}{2}$.

This theorem was proved in a different way by A. J. Hoffman and W. Jacobs in [86]. A similar theorem, for the case where the r_i first increase and then decrease, is contained in [19].

8–6 Buying, storing, selling. Let it be known that r_i units of some commodity will be required in the ith month $(i = 1, \ldots, n)$ and that the cost of production in the ith month is c_i per unit, while storing the production of the ith month until it is required during the jth month costs d_{ij} per unit. There is also an upper limit of L_i units on the production of the ith month. How much should be produced in each month, so that the total of the production and storage costs is minimized?

This problem can be formulated as a Transportation Problem with the following combined requirement and cost table:

	r_1	r_2	$\cdots$	r_n	$\sum_i (L_i - r_i)$
L_1	c_1	$c_1 + d_{12}$		$c_1 + d_{1n}$	0
L_2		c_2		$c_2 + d_{2n}$	0
.					
.					
.					
L_n				c_n	0

The cost in the cells of the lower left-hand triangle, which do not have any realistic meaning (you cannot produce for a time which has already passed), is a very high number. In order to make this scheme feasible, we must have

$$\sum_{i=1}^{t} (L_i - r_i) \geqslant 0 \qquad \text{for all } t = 1, \ldots, n.$$

(Transformation into a slightly different type of Transportation Problem is shown in A. S. Manne [102].)

It has been pointed out by S. Johnson in [91] that such a problem can be solved by inspection. The requirements are met, one after the other, by assigning to them the cheapest production still available when they are considered.

EXAMPLE 8–3. (Cf. [15], pp. 13 ff.) The table

	5	6	8	6	11
			Costs		
9	1	2	3	4	0
9		4	5	6	0
9			2	3	0
9				4	0

is solved by

	5	6	8	6	11
9	5	4			
9		2			7
9			8	1	
9				5	4.

The very obvious proof of the validity of this procedure is left to the reader.

When there are capacity restrictions on storage as well, the same procedure is available with slight modifications, as shown in [29]. A similar procedure can also be used to solve the following problem: There is a capacity limitation on, but no cost connected with storage. The cost per unit of production is c_i as before, and the selling price per unit in the ith month is p_i $(i = 1, \ldots, n)$. In each period selling takes place before buying. E. M. L. Beale, in [26], has developed the following procedure.

At each stage we define a tentative buying period i_t, a tentative selling period j_t, a firm buying period i_f, and a firm selling period j_f. The start of the procedure depends on the storage in the warehouse.

If the warehouse is originally full, we start with $i_t = 0$, and if it is empty, we start with $i_t = 1$. If the initial storage is A, and the storage capacity is B, then we imagine two warehouses, one with capacity A which is full, and one with capacity $B - A$ which is empty.

Whenever a value of i_t has been determined, look for the first j, larger than i_t, such that p_j is larger than the smallest c_i for all i from i_t (inclusive) to j (exclusive). Denote that j by j_t, and the subscript of that smallest c_i by i_f.

Then look for the first i not smaller than j_t such that c_i is smaller than the largest p_j for all j from j_t (inclusive) to i (inclusive). Let that i be the new i_t, and the subscript of that largest p_j is j_f. Repeat the steps until the highest i or j is reached.

The optimal policy consists in selling the entire stock in all periods j_f, and filling it up in all periods i_f.

S. E. Dreyfus, in [59], has indicated a simple algorithm based on Dynamic Programming.

The Warehouse Problem was first mentioned by A. S. Cahn in [33].

8–7 Race-course betting. Suppose that n horses (or greyhounds) have been entered in a race and that you are offered odds a_i on horse i, which means that if the ith horse wins, then you receive a_i times your stake as a win, and you get your stake back, while if that horse does not win, then you lose your stake.

You ask yourself how to distribute your stakes so that the smallest amount you could get as a result of the race is as large as possible. You will also want to know whether that amount will be positive, since otherwise there would not be any point in this exercise.

Denote the portions of your total stake on the several horses by $x_1, \ldots, x_n$. Then $x_1 + \ldots + x_n = 1$. If horse i wins, then you win, on balance,

$$-x_1 - \ldots + a_i x_i - \ldots - x_n.$$

Introducing $c_i = a_i + 1$, this can be written $c_i x_i - 1$. If we denote the smallest of these values by v, then we have the constraints

$$x_1 + \ldots + x_n = 1,$$
$$c_i x_i - 1 \geqslant v \qquad (i = 1, \ldots, n).$$

$$\text{Maximize } v.$$

The unknown v is not sign-restricted and we may therefore eliminate it. Introducing slack variables y_i, the result can be written

$$c_i x_i - y_i = c_1 x_1 - y_1 \qquad (i = 2, \ldots, n),$$
$$x_1 + \ldots + x_n = 1.$$

$$\text{Maximize } v = \frac{\sum_i c_i x_i - \sum_i y_i - n}{n}.$$

There are now n constraints with $2n$ variables. A basic feasible solution is

$$x_i = \frac{k}{c_i} \qquad \text{where } k = 1 / \sum_i \left(\frac{1}{c_i} \right),$$
$$y_i = 0, \qquad (i = 1, \ldots, n).$$

Solving for the basic variables x_i, we obtain

$$x_i = \frac{y_i}{c_i} + \frac{k}{c_i} \left[1 - \sum_j \frac{y_j}{c_j} \right],$$
$$v = k - 1 - k \sum_j \frac{y_j}{c_j}.$$

Since the c_j are positive, this is the optimal solution, and the optimal value of B is $k - 1$.

The answer to the betting problem is therefore that the stakes should be distributed in proportions $1/(a_i + 1)$. Then the original constraints are all satisfied as equations, so that whichever horse wins, the pay-off is the same, viz. $k - 1$.

Of course, this is only worth undertaking if $k - 1$ is positive, that is if

$$\sum_i \frac{1}{a_i + 1} < 1.$$

EXAMPLE 8-4. Let the odds offered be $3:1$, $7:1$, $3:2$, $1:1$. The best distribution of stakes is then in proportions $1/4:1/8:2/5:1/2$. The sum of these fractions is 1.275. Don't touch it!

8-8 A game problem. Two players A and B play the following game: Either player chooses an integer from the numbers one to five. If the total of their choices is even, then A gets that total from B; if it is odd, then B gets the total from A. The following *pay-off* table indicates how the outcomes depend on the choices of the players. The entries in the table are the amounts that A wins.

		\multicolumn{5}{c}{B chooses}
		1	2	3	4	5
	1	2	−3	4	−5	6
	2	−3	4	−5	6	−7
A chooses	3	4	−5	6	−7	8
	4	−5	6	−7	8	−9
	5	6	−7	8	−9	10 .

In the spirit of von Neumann's *Theory of Games* (see, for instance, [16]) each player should choose his number according to chance, and in particular with such probabilities that the smallest expected value of his win (which depends on the opponent's choice as well) should be as large as possible.

Let the probability with which A chooses i be p_i, so that

$$p_1 + p_2 + p_3 + p_4 + p_5 = 1. \tag{1}$$

If the smallest expected value is v, then we have

$$2p_1 - 3p_2 + 4p_3 - 5p_4 + 6p_5 \geqslant v,$$

$$\vdots \tag{2}$$

$$6p_1 - 7p_2 + 8p_3 - 9p_4 + 10p_5 \geqslant v,$$
$$p_i \geqslant 0 \quad (\text{all } i), \tag{3}$$

and v is to be maximized.

For reasons which will appear presently, we add ten times the equation (1) to the five inequalities (2). This gives

$$12p_1 + 7p_2 + 14p_3 + 5p_4 + 16p_5 \geqslant v + 10,$$
$$7p_1 + 14p_2 + 5p_3 + 16p_4 + 3p_5 \geqslant v + 10,$$
$$14p_1 + 5p_2 + 16p_3 + 3p_4 + 18p_5 \geqslant v + 10,$$
$$5p_1 + 16p_2 + 3p_3 + 18p_4 + p_5 \geqslant v + 10,$$
$$16p_1 + 3p_2 + 18p_3 + p_4 + 20p_5 \geqslant v + 10.$$

Maximizing v is equivalent to maximizing $v + 10 = w$, say. Now all coefficients on the left-hand side are positive, so that we can be certain that the maximal w will be positive. Introducing $p_i/w = x_i$, we have

$$12x_1 + 7x_2 + 14x_3 + 5x_4 + 16x_5 \geqslant 1,$$
$$\cdot$$
$$\cdot$$
$$\cdot$$
$$16x_1 + 3x_2 + 18x_3 + x_4 + 20x_5 \geqslant 1,$$
$$x_i \geqslant 0 \quad \text{(all } i\text{)}.$$
$$\text{Minimize } x_1 + x_2 + x_3 + x_4 + x_5.$$

Similarly, remembering that B wins the negative of the entries in the pay-off table, his problem can be written

$$12y_1 + 7y_2 + 14y_3 + 5y_4 + 16y_5 \leqslant 1,$$
$$\cdot$$
$$\cdot$$
$$\cdot$$
$$16y_1 + 3y_2 + 18y_3 + y_4 + 20y_5 \leqslant 1,$$
$$y_j \geqslant 0 \quad \text{(all } j\text{)}.$$
$$\text{Maximize } y_1 + y_2 + y_3 + y_4 + y_5.$$

The two problems are dual to one another. The resulting problems would have turned out to be dual to one another even if the pay-off table had not been symmetrical to begin with. Therefore by solving one of the linear programming problems we solve implicitly the other as well.

The problems have been solved in Exercise 5–12. The optimal value of the objective function is 0.1; hence the optimal values of the p_i are 10 times the corresponding x_i and the optimal value of v, the *value of the game*, is $10 - 10 = 0$. This means that if both players choose their numbers according to the optimal p_i, then neither will win or lose in the long run.

The optimal p_i are as follows:

	p_1	p_2	p_3	p_4	p_5
	3/8	1/2			1/8
or	1/4	1/2	1/4		
or		1/4	1/2	1/4	
or			1/4	1/2	1/4.

If both players have only 1 to 4, or 1 to 3, or 1, 2 to choose from, then the game's problem is solved by (c) of Exercise 5–12. It will be seen that in the last case A loses in the long run 1/12.

CHAPTER 9

PARAMETRIC LINEAR PROGRAMMING

Up to now we have always assumed that the constants in the constraints as well as in the objective function were known. However, in applications of our techniques it will be frequently found that the greatest difficulty rests precisely in finding the values of these constants. They depend, in industrial and technical applications, on the technology considered and we call them, therefore, *technological coefficients*.

The question of how to ascertain these values is, of course, not mathematical. But the mathematician must be concerned with how the uncertainty of the results depends on the uncertainty of the data. We shall be dealing with this problem again in Chapter 11 (stochastic programming); we deal here with one aspect of this problem that can be isolated from others and studied with relative ease.

Let us assume that the constants in the objective function depend on a single parameter in the following way. Consider two sets of coefficients in the objective function

$$c_1 x_1 + \ldots + c_N x_N, \quad \text{say } c_1', \ldots, c_N' \text{ and } c_1'', \ldots, c_N''.$$

We are interested in all sets

$$c_1' + t c_1'', \ldots, c_N' + t c_N'',$$

where the parameter t can take any finite value.

To begin with, imagine that t has a fixed value, t_0. Applying the Simplex method it is easy to establish which values of t lead to the same set of basic variables as t_0 does. The values of the variables depend, of course, on the set of basic variables, but not on the value of t, which appears only in the objective function.

We discuss these ideas now in greater detail. We assume that the constraints are consistent—otherwise the problem of parameters in the objective function is irrelevant. Write them

$$a_{1j} x_1 + \ldots + a_{Nj} x_N = b_j \quad (j = 1, \ldots, m),$$

and let the objective function, to be minimized, be

$$c_1 x_1 + \ldots + c_N x_N.$$

First, we consider the coefficients

$$c_i = c_i' + t_0 c_i''.$$

187

It is possible that the objective function has then no lower bound, and this will emerge during the Simplex procedure in the usual way. There will exist a value

$$z_{0h} = z'_{0h} + t_0 z''_{0h} > 0,$$

while no value $z_{ih} > 0$ appears in the column of x_h that could be used as a pivot. If $z''_{0h} = 0$, then this remains true for any value of t. Otherwise it holds for all t for which $z'_{0h} + t z''_{0h} > 0$. By our assumption this holds for $t = t_0$, and therefore, when $z''_{0h} > 0$, then any $t > t_0$ will give an objective function without lower bound, and if $z''_{0h} < 0$, then this is the case for $t < t_0$.

Geometrically, the choice of t means the choice of a preferred direction. If there are any constraints at all, there must be some direction in which the feasible region is bounded. Assume that this is the case for $t = t_1 \geqslant t_0$. We explore the situation for $t > t_1$. The details for the case $t < t_1$ can be found in the same way and need not be discussed here.

Let, for $t = t_1$, the minimum of the objective function be C_0, and let the corresponding values of the variables be $x_1^0, \ldots, x_N^0$, where the first m are basic. The last row in the final Simplex tableau is then

$$z'_{0.m+1} + t_1 z''_{0m+1} \quad \cdots \quad z'_{0N} + t_1 z''_{0N} \quad C_0.$$

If all z''_{0j} are zero, or negative, then t can be increased without bounds, and the tableau remains optimal. Otherwise it remains optimal only until t reaches the smallest of the ratios $-z'_{0j}/z''_{0j}$ for positive z''_{0j}. Any increase beyond this value of t destroys the optimality of the tableau. No higher value of t can make the tableau optimal again.

We consider now these two cases: (i) t can be increased (to a suitable extent) from t_1 without destroying the optimality of the present tableau, and (ii) any increase of t destroys this optimality.

(i) In this case t can be increased to the smallest of the ratios $-z'_{0j}/z''_{0j}$, considering only those j for which the denominator is positive. When t reaches this value, say t_2, then the objective function will have a zero coefficient of one of the nonbasic variables, say of x_{m+1}. After an exchange of variables, the numerical values in the new restricted tableau are the same as before (because t_2 makes $z_{0h} = 0$), though they do not refer any more to precisely the same nonbasic variables. This is thus again an optimal tableau, but now t_2 is the smallest value for which this is true. We can repeat our argument and find a succession of values of t at which the optimal set of basic variables changes.

(ii) After performing an exchange as in (i), it is possible that only the value of t just reached makes the new tableau optimal. But this cannot go on for ever, because only a finite number of sets of basic variables can exist, and one of them must appear when we increase t by an arbitrarily small amount. We can thus carry on and find the values of t where a change occurs. These values will necessarily be finite in number.

EXAMPLE 9–1. We choose an example where the objective function is to be maximized. It will illustrate how the argument above must be applied to this case. Maximize

$$(3 - t)x_1 + (6 - t)x_2 + (2 - t)x_3,$$

subject to

$$3x_1 + 4x_2 + x_3 \leqslant 2,$$
$$x_1 + 3x_2 + 2x_3 \leqslant 1.$$

The first tableau is

	$3 - t$	$6 - t$	$2 - t$	
	x_1	x_2	x_3	
x_4	3	4	1	2
x_5	1	3*	2	1
B	$t - 3$	$t - 6$	$t - 2$	0

This is optimal as long as $t \geqslant 6$. When $t = 6$, then we can exchange x_2 and x_5:

		$3 - t$		$2 - t$	
		x_1	x_5	x_3	
	x_4	5/3*	$-4/3$	$-5/3$	2/3
$6 - t$	x_2	1/3	1/3	2/3	1/3
	B	$2t/3 - 1$	$2 - t/3$	$t/3 + 2$	$2 - t/3$

This is optimal for t between $3/2$ and 6, both limits included. When t is at $3/2$, we can exchange x_1 and x_4:

				$2 - t$	
		x_4	x_5	x_3	
$3 - t$	x_1	3/5	$-4/5$	-1	2/5
$6 - t$	x_2	$-1/5$	3/5	1	1/5
	B	$3/5 - 2t/5$	$t/5 + 6/5$	$t + 1$	$12/5 - 3t/5$

This is optimal for t between -1 and $3/2$. It should be noted that $t = 0$ lies within this range, and the answer is then the same as that found in Example 5-1, as it must be. When $t = -1$, we can exchange x_2 and x_3:

$$6 - t$$

		x_4	x_5	x_2	
$3-t$	x_1	$2/5$	$-1/5$	1	$3/5$
$2-t$	x_3	$-1/5$	$3/5$	1	$1/5$

B	$4/5 - t/5$	$3/5 - 2t/5$	$-1 - t$	$11/5 - 4t/5$

This is optimal, provided t does not exceed -1.

To summarize, we have the same optimal values of the variables within each of the following ranges of the parameter:

below -1, -1 to $3/2$, $3/2$ to 6,   above 6.

The presentation in this chapter is based on [73, 110].

EXERCISES

9-1. Find the various basic sets and the values of the objective function corresponding to the values of the parameter t in the following problem:

$$x_1 + x_2 + x_3 = 3,$$
$$x_1 - 2x_2 + x_4 = 1,$$
$$-2x_1 + x_2 + x_5 = 2.$$

Maximize $(1 + t)x_1 - (1 - t)x_2$.

9-2. Find the various basic sets and the values of the objective function corresponding to the values of the parameters u and v in the following problem:

$$x_1 + x_2 \leqslant 3, \qquad 2x_1 - x_2 \leqslant 2, \qquad x_1, x_2 \geqslant 0.$$

Maximize $(2 + 2u + v)x_1 + (1 + u - v)x_2$.

9-3. Solve the following problem for all values of t:

$$3x_1 + x_4 + 2x_5 = 12,$$
$$3x_2 - x_4 + x_5 + y = 3,$$
$$x_3 + x_4 + x_5 = 9.$$

Minimize $x_2 - x_1 + ty$.

9-4. Solve the following problem for all values of t:

$$-2x_1 + 2x_2 + x_3 \leqslant 12,$$
$$3x_1 - 18x_2 - 4x_3 \leqslant 24,$$
$$x_1 + 2x_2 + 4x_3 \leqslant 24.$$

Minimize $C = (-1 + t)x_1 + (2 + t)x_2 + (2 + t)x_3$.

CHAPTER 10

DISCRETE LINEAR PROGRAMMING

10–1 Travelling salesman. The *travelling salesman* problem, which has been discussed in a number of papers [51, 52, 64], is concerned with the following topic. Let n towns be given, with all the distances between any two of them. It is required to start from a given town, to visit all the others, and to return to the starting point, in such a way that the total distance travelled is as short as possible. There are, of course, $(n-1)!$ possible circular tours, and the problem is the same, whatever the starting point. The linear programming formulation of this problem can be achieved in various ways.

Let the variable x_{ijt} equal unity, if the tth directed arc is from town i to town j, and zero otherwise. The subscripts i, j, and t can take all integer values from 1 to n. The variables are subject to the following constraints: For all values of t, some arc must be traversed, hence

$$\sum_i \sum_j x_{ijt} = 1 \qquad \text{for all } t.$$

For all towns, there is just one other town which is being reached from it, at some time, hence

$$\sum_j \sum_t x_{ijt} = 1 \qquad \text{for all } i.$$

For all towns, there is some other town from which it is being reached, at some time, hence

$$\sum_i \sum_t x_{ijt} = 1 \qquad \text{for all } j.$$

When a town is reached at time t, it must be left at time $t+1$ (or at time 1 if $t = n$); hence

$$\sum_i x_{ijt} = \sum_k x_{jkt+1} \qquad \text{for all } j \text{ and } t.$$

This excludes disconnected circular tours.

We want to minimize $\sum_i \sum_j \sum_t d_{ij} x_{ijt}$, where d_{ij} is the distance from i to j. It is not necessarily symmetric in the subscripts. A one-way street is an obvious counter-example.

In addition to these constraints, x_{ijt} can only have the value 1 or 0. We may express this by saying that $0 \leqslant x_{ijt} \leqslant 1$, and that the variable can only have integer values.

When we dealt with the transportation problem, we saw that provided the marginal totals were integers all basic feasible solutions were also expressed in integers. In the travelling salesman problem this is not automatically true, although the formulation is similar to that in the transportation problem. The following example (due to E. M. L. Beale) shows the possible complications: Let five towns be given, A, B, C, D, and E, and let their positions and distances be as indicated in Fig. 10-1 (in a realistic problem the distances, or times taken between two towns, do not have to satisfy even the triangular inequality).

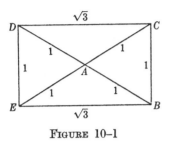

FIGURE 10-1

The following values of the variables satisfy the constraints

$$x_{AB1} = x_{BC2} = x_{CB3} = x_{BC4} = x_{CA5} = \tfrac{1}{2},$$

and also

$$x_{AD1} = x_{DE2} = x_{ED3} = x_{DE4} = x_{EA5} = \tfrac{1}{2}.$$

But this is impossible to obey. It would mean that half a salesman travels from A to B, then to C and back to B, again to C, and finally reverting to A, while the other half moves from A to D, oscillates between D and E and D and E, and then returns to A. The total path of this would be 5, and we see from the figure that it is (one of) the shortest possible, because no arc of greater length than 1 is used. We see also from the figure that any genuine circuit will have to include such a longer arc, and its total length will be at least $4 + \sqrt{3}$. Thus, without making some effective provision for the variables to have only integer values, we obtain a shortest circuit without realistic meaning.

10-2 Allocation problem. We turn to another example.

EXAMPLE 10-1. Figure 10-2 shows the position of three television stations. Due to their power and geographic location, some regions overlap, as shown, and any two stations with overlapping regions must have different frequency allocations. There are two different wavelengths at our disposal. How should they be allocated?

First, we formalize the conditions about overlap. Let $s_{ij} = 1$ if there is no overlap between station i and station j, and $s_{ij} = 0$

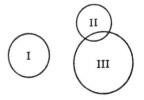

FIGURE 10-2

if there is. In the present case, we have the following table of the s_{ij}:

	I	II	III
I	0	1	1
II	1	0	0
III	1	0	0 .

Let x_{ik} indicate whether station i has frequency k allocated to it (in which case $x_{ik} = 1$) or not ($x_{ik} = 0$). The conditions to which they are subject are: Each station has precisely one frequency allocated to it; hence

$$x_{11} + x_{12} = 1,$$
$$x_{21} + x_{22} = 1,$$
$$x_{31} + x_{32} = 1.$$

The overlap condition is

$$x_{ik} + x_{jk} \leqslant 1 + s_{ij},$$

because then the two variables cannot both be unity (and one of them must therefore be zero) unless $s_{ij} = 1$, when there is no overlap.

In this problem no optimization arises. The conditions are satisfied, for instance, by either of the following allocations:

I	II	III		I	II	III
1	1	2	or	1	2	1.

Let us choose the second of these. Now a fourth station is being built, and the overlap conditions are as shown in Fig. 10–3. It is now not possible to allocate to IV either frequency 1 or 2. (Had we taken the other solution, then frequency 2 could have been given to the fourth station.) We ask how all frequencies could be allocated in such a way that the smallest number of the stations previously in existence is being disturbed.

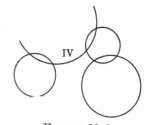

FIGURE 10–3

This leads again to a discrete linear programming problem. Imagine that we want to solve the second problem without reference to the first. We use now notation y_{ik} instead of x_{ik} and it is easy to write down the constraints for the new case. At this stage the requirement of minimum disturbance supplies an objective function. We want to maximize

$$\sum_i \sum_k x_{ik} y_{ik} \quad \text{(the number of undisturbed stations)}.$$

In the present example this number is

$$y_{11} + y_{22} + y_{31},$$

and leads to

$$y_{11} = y_{21} = y_{32} = y_{42} = 1,$$

and the other unknowns zero. The allocation is that suggested by the first solution of the original problem, with frequency 2 allocated to the new station.*

10–3 Logical relations. In the previous section we have introduced variables which must be either 0 or 1. The same device can be used for other logical relationships as well. For instance, the fact that a variable x has either value a_1 or a_2, . . ., or a_n may be written

$$x = d_1 a_1 + \ldots + d_n a_n,$$

with the added conditions that

$$d_1 + \ldots + d_n = 1 \quad \text{and all } d_i \text{ either 0 or 1.}$$

If the a_i are 1, 2, . . ., n, then it is even simpler to write $x = d_1 + \ldots + d_n$, where all d_i are either 0 or 1 as before, but do not necessarily add up to 1. (This remark is due to H. Markowitz and A. Manne [104].)

More complicated logical relationships can be formulated in a similar fashion. For instance, let it be required that of the following p groups of inequalities at least k should hold; it should be understood that we say that a group of inequalities holds if all its members are satisfied.

$$\text{Group 1: } f_{11}(x) \geqslant 0, \ldots, f_{1n_1}(x) \geqslant 0,$$

.

.

.

$$\text{Group } p: f_{p1}(x) \geqslant 0, \ldots, f_{pn_p}(x) \geqslant 0,$$

where the $f_{ij}(x)$ are functions of the set $x = (x_1, \ldots, x_n)$.

We assume that we know lower bounds for the values of these functions; thus

$$f_{ij}(x) \geqslant L_{ij}(x).$$

We can then write

$$f_{1j}(x) \geqslant d_1 L_{1j}(x) \qquad \text{for } j = 1, \ldots, n_1,$$

.

.

.

$$f_{pj}(x) \geqslant d_p L_{pj}(x) \qquad \text{for } j = 1, \ldots, n_p,$$
$$d_1 + \ldots + d_p = p - k \quad \text{and all } d_i \text{ either 0 or 1.}$$

_____

* This method of solving the problem emerged in discussions with Dr. A. J. Hoffman and Mr. J. A. P. Deutsch.

Indeed, $p - k$ of the d_i will be unity, and the other k will be zero. The latter produce k groups of conditions which state that

$$f_{ij}(x) \geqslant 0 \qquad \text{for all } j = 1, \ldots, n_i.$$

The remaining d_i lead to inequalities which are automatically satisfied by virtue of the definition of the $L_{ij}(x)$, and thus they do not add any further restriction. For these and similar ideas, see [45].

We mention a few special cases of this method, concerning linear functions $f_{ij}(x_1, \ldots, x_n)$.

EXAMPLE 10–2. Let $p = 2$ and $k = 1$. This means that the inequalities of at least one of two groups must hold. For instance, let the two groups consist of

First group		Second group	
$x_2 \geqslant 0,$	(a)	$-x_2 + 1/\sqrt{3} \geqslant 0,$	(d)
$\sqrt{3}x_1 - x_2 \geqslant 0,$	(b)	$3x_1 + \sqrt{3}x_2 - 1 \geqslant 0,$	(e)
$-\sqrt{3}x_1 - x_2 + \sqrt{3} \geqslant 0.$	(c)	$-\sqrt{3}x_1 + x_2 + 2/\sqrt{3} \geqslant 0.$	(f)

The points whose coordinates satisfy at least one group of inequalities form the interior of a Star of David (Fig. 10–4).

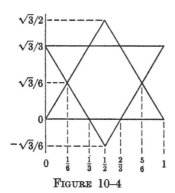

FIGURE 10–4

For the formulations which we have in mind, we must have lower bounds for all the functions. If one group of constraints is satisfied, then we do not add anything new if we demand that the value of any of the left-hand side of the other group should not exceed that value which it assumes when the coordinates of that vertex of the first triangle are inserted, which lies on the negative side of it. By this argument we obtain the following lower bounds:

First group	Second group
(a) $-\sqrt{3}/6$	(d) $-\sqrt{3}/6$
(b) $-\sqrt{3}/3$	(e) -1
(c) $-\sqrt{3}/3$	(f) $-1/\sqrt{3}$.

10–4 Fixed charge problem. W. M. Hirsch and G. B. Dantzig have studied an analytic problem in [84]. They consider the case of activities i which have a fixed setup cost s_i attached to them, but only if they are carried out at a level $x_i > 0$, in addition to a cost $c_i x_i$ proportional to that level. The objective function is thus

$$\sum_i c_i x_i + \sum_i s_i d(x_i),$$

where $d(x_i)$ is unity for $x_i > 0$, and zero otherwise.

Consider the special case where all s_i are equal, say s_0, and assume that the set of m constraints is not degenerate, so that any solution contains at least m variables with positive values. Then the minimum of the objective function is obtained for precisely those values x_i^0 of the variables x_i which minimize the simpler function $\sum_i c_i x_i$. The proof is contained in the following sequal of relations:

$$\sum_i c_i x_i^0 + \sum_i s_0 d(x_i^0) = \sum_i c_i x_i^0 + m s_0 \geqslant \min\left[\sum_i c_i x_i + s_0 \sum_i d(x_i)\right]$$

$$\geqslant \min \sum_i c_i x_i + s_0 \min \sum_i d(x_i)$$

$$\geqslant \min \sum_i c_i x_i + m s_0$$

$$= \sum_i c_i x_i^0 + m s_0.$$

The two extreme terms are the same, and they are therefore also equal to one of the middle terms

$$\min\left[\sum_i c_i x_i + s_0 \sum_i d(x_i)\right]. \qquad \text{Q.E.D.}$$

10–5 Discrete linear programming algorithms. We shall introduce methods to deal with *discrete linear programming* problems, i.e. linear programming problems for which it is also required that variables take integer values only. Such problems belong to two distinct categories: either all variables, or a selected set of them, must be integers. We deal first with the former case.

The principle which will guide us, and which has been used in other contexts as well, is that of introducing *cuts*. These are constraints which exclude some of the other feasible solutions but without excluding any solutions with certain variables having integer values.

One rather natural cut of this type has been suggested by G. B. Dantzig [45].

Whenever the set of nonbasic variables is known, the values of the basic variables, and that of the objective function, are also given. If the values of the variables are not all integers, then some other set of nonbasic variables is required, and at least one of these which are now nonbasic will then acquire a value equal to or larger than 1. Hence the further constraint, stating that the sum of the present nonbasic variables is at least 1, can be introduced. In some cases this principle leads to the final solution, but it is not known whether such a procedure must terminate after a finite number of steps.

In the method due to R. E. Gomory (see [75]), which will now be described, we solve, first, the problem without taking any account of the requirement of integrality. If all variables in the final solution have integer values, then we have finished. If this is not so, then we add a further constraint.

Assume that the final tableau in the Simplex method is

	x_{m+1}	x_N	
x_1	$z_{1,m+1}$	z_{1N}	x_1^0
.			
.			
.			
x_m	$z_{m,m+1}$	z_{mN}	x_m^0
	$z_{0,m+1}$	z_{0N}	z_{00}

and assume that x_1^0 is not an integer.

We split every constant in the row of x_1 into its largest lower integer and a non-negative fraction; thus

$$z_{1j} = N_{1j} + f_{1j}, \qquad x_1^0 = N_{10} + f_{10},$$

where

$$0 \leqslant f_{1j} < 1 \quad \text{for } j = m+1, \ldots, N,$$

and

$$0 < f_{10} < 1.$$

Define

$$k = f_{1m+1}x_{m+1} + \ldots + f_{1N}x_N \geqslant 0.$$

Subtract this from

$$N_{10} + f_{10} = x_1 + (N_{1m+1} + f_{1m+1})x_{m+1} + \ldots + (N_{1N} + f_{1N})x_N.$$

We obtain (transferring N_{10} to the right-hand side)

$$f_{10} - k = x_1 + N_{1m+1}x_{m+1} + \ldots + N_{1N}x_N - N_{10}.$$

The condition that all variables have integer values implies that the right-hand side must also be integer, and it is not larger than f_{10}. But the latter is a fraction, and hence the right-hand side cannot exceed zero. It follows that the requirement of integrality implies that $k \geqslant f_{10}$.

From the definition of k, we can write

$$x_{N+1} - f_{1m+1}x_{m+1} - \ldots - f_{1N}x_N = -f_{10},$$

where x_{N+1} is a new variable, which must have non-negative values and must be an integer, since it is identical with $k - f_{10}$. This equation is the *cut* which we introduce, and we see that the resulting problem is again one of discrete linear programming. At the stage represented by the final tableau of the Simplex method, which we have just reached, the value of x_{N+1} is $-f_{10}$, hence negative and fractional. We add the following row to the Simplex tableau

$$x_{N+1} \qquad -f_{1m+1} \qquad\qquad -f_{1N} \qquad -f_{10}.$$

The objective function has coefficients with those signs that are required in the final tableau. Hence we can apply the Dual Simplex method to make x_{N+1} nonbasic and therefore zero. Possibly now all variables have integer, non-negative values, in which case we have finished. If this is not so, then we carry on with the Dual Simplex method, introducing new variables where required.

Because we introduce a new variable whenever we want to remove a fraction, it is not immediately obvious that a final answer is reached after a finite number of steps. R. Gomory has proved that this is true for the case when it is known that the objective function has a lower as well as an upper bound, but the proof is too lengthy to be reproduced here.

If we multiply an equation by a number, t say, then we might get other fractions, and thus other further constraints. We obtain altogether D different cases if D is the least common denominator of the constants of the equation, because multiplication by t and by $D + t$ produces the same fractional parts. In particular, if $t = D - 1$, then we have

$$(D - 1)x_1 + (D - 1)z_{1m+1}x_{m+1} + \ldots + (D - 1)z_{1N}x_N = (D - 1)x_1^0.$$

Split again into integers and fractions, writing

$$(D - 1)z_{1j} = N'_{1j} + f'_{1j}, \qquad (D - 1)x_1^0 = N'_{10} + f'_{10},$$

i.e.

$$Dz_{1j} - N_{1j} - f_{1j} = N'_{1j} + f'_{1j},$$

so that

$$f_{1j} + f'_{1j} = 1,$$

unless both are zero. This holds for $j = 0, m + 1, \ldots, N$.

EXAMPLE 10–3. Given

$$-x_1 + 10x_2 \leqslant 40,$$
$$x_1 + x_2 \leqslant 20.$$

Maximize $-10x_1 + 111x_2 = B$, say, x_1 and x_2 to be non-negative integers.

The Simplex method produces the following final tableau:

		x_3	x_4	
−10	x_1	−1/11	10/11	$14\frac{6}{11}$
111	x_2	1/11	1/11	$5\frac{5}{11}$
	B	11	1	460

The variables have fractional values. We choose to fix our attention on x_2, and add the further row

$$x_5 \quad -1/11 \quad -1/11 \quad -5/11.$$

One application of the Dual Simplex method gives

		x_3	x_5	
−10	x_1	−1	10	10
111	x_2	0	1	5
	x_4	1	−11	5
	B	10	11	455

This contains the answer.

It should be noticed that the values of the variables maximizing B are not those which we would have obtained by rounding the values in the previous tableau to the nearest integers, up or down, even if we had taken precautions not to obtain thereby a point which is not feasible, e.g. (14, 6). We could also have added the further row

$$x_5' \qquad -10/11 \qquad -10/11 \qquad -6/11.$$

This would have needed two further iterations, and altogether two new variables. Furthermore, we could have attended to x_1. There is not enough known, as yet, to be able to say which choice is the best in all circumstances.

It is of interest to consider a geometric representation of this method (Fig. 10–5). The final tableau above for the variables x_1, x_2, x_3, and x_4 indicates the following relationship:

$$-x_3 + 10x_4 \leqslant 160,$$
$$x_3 + \quad x_4 \leqslant 60,$$
$$B = 460 - 11x_3 - x_4.$$

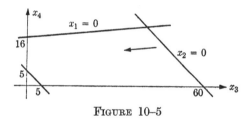

FIGURE 10–5

The addition of the new non-negative variable means the addition of a line parallel to $x_2 = 0$, which cuts off a corner of the previously feasible region, so that the final answer ($x_3 = 0$, $x_4 = 5$) is now a vertex.

10–6 Discrete linear programming alogarithms. Mixed case. The method which we have described is only applicable when all variables must have integer values. If this requirement applies only to some "specified" variables, a method due to A. H. Land and A. Doig can be used [57]. This method is, of course, also available when all variables are "specified".

We start again with the Simplex method to obtain the optimum without regard for integrality, and then trace our steps back to a point which satisfies all requirements.

For the sake of our argument we shall assume that we want to maximize. It will be quite clear how to apply our argument in the opposite case. Also, when we mention constraints, we shall not include in this expression the requirement of integrality.

Given any value of the objective function, B, we can determine the range of values of any specified variable, say x_1, compatible with that value of B, and with the constraints. The region defined by the possible values of B and the possible values of x_1 which produce that value in conjunction with appropriate values of the remaining variables, is convex, since given any two values of B and two values of x_1 compatible with them, respectively, any intermediate value of B can be produced by a suitable linear combination of the values of x_1.

If B takes its largest value, there will, in general, only be one value of x_1 compatible with it. This value can be defined by linear programming.

Our procedure can be described as follows: We maximize B, subject to the constraints only.

Let x_1 be one of the specified variables. If it is not already integral, then we know that B must be reduced. It follows from the convexity of the region (x_1, B) just mentioned that if x_1 is the only specified variable, then it is only necessary to reduce B either to a value for which the minimum of x_1 is the nearest lower integer, or to a value for which the maximum of x_1 is the nearest higher integer. When we first reach either of these values, we stop.

If x_1 is not the only specified variable, then we keep it fixed at its nearest lower value and deal in the same way with another specified variable; also, we do this keeping x_1 fixed at its nearest higher value, and so on. At each step when we have reached integer values of some specified variables, we note the value of B and branch out from all these points, until there is no other specified variable left. Of all such solutions we choose finally that corresponding to the highest value of B.

We cannot enter here into all the computational details, but we discuss a few points in connection with an example.

EXAMPLE 10–4. We choose the problem discussed in Example 10–3. It is sufficient to specify x_1 and x_2. The others will then be integral by virtue of the constraints which define them. The tableau

	x_3	x_4	
x_1	$-1/11$	$10/11$	$14\frac{6}{11}$
x_2	$1/11$	$1/11$	$5\frac{5}{11}$
B	11	1	460

would be optimal but for the fractional values of the variables. We want to reduce B in such a way that the minimum of x_1 for the reduced value is 14. The relations between the variables and B must remain the same, so that we can write

$$x_1 + z_{13}x_3 + z_{14}x_4 = 14\tfrac{6}{11},$$
$$x_2 + z_{23}x_3 + z_{24}x_4 = 5\tfrac{5}{11},$$
$$B + t_1 - 460 + z_{03}x_3 + z_{04}x_4 - t_1 = 0.$$

Consider the first t_1 in the last equation to be the reduction of B which must be accepted if x_1 is to be 14, and the second t_1 as a new variable with non-negative values. In tableau form we write

		x_3	x_4	t_1	
	x_2	$1/11$	$1/11$	0	$5\tfrac{5}{11}$
$H_1 = B + t_1 - 460$		11	1^*	-1	0
	x_1	$-1/11$	$10/11$	0	$14\tfrac{6}{11}$

x_1 is the new objective function to be minimized. It is clear how this tableau is obtained from the previous one.

To minimize x_1 we apply a transformation and obtain

	x_3	H_1	t_1	
x_2	$-10/11$	$-1/11$	$1/11$	$5\tfrac{5}{11}$
x_4	11	1	-1	0
x_1	$-10\tfrac{1}{11}$	$-10/11$	$10/11$	$14\tfrac{6}{11}$

Here t_1 has some non-negative value, but the value of x_1 is minimized, in view of the signs of the bottom values in the x_3 and the H_1 columns. We choose t_1 so that

$$x_1 + 10t_1/11 = 14\tfrac{6}{11} \qquad \text{while } x_1 = 14.$$

Hence

$$t = \tfrac{3}{5},$$

and we have

$$x_1 = 14, \qquad x_2 = 5\tfrac{5}{11} - t_1/11 = 5\tfrac{2}{5}.$$

Also

$$H_1 = B + t_1 - 460 = 0,$$

and hence

$$B = 459.4.$$

This is the way to find the value of the objective function B which makes the minimum of all compatible values of x_1 equal to 14. The value of x_2 is still fractional, and we must find, *inter alia*, that value of B for which the minimum of x_2 is 5, while x_1 retains its value 14.

The latter requirement can be satisfied by introducing a further variable, say F, such that

$$x_1 - F = 14,$$

and stipulating that F must always remain at zero. This, together with the information contained in the last tableau (with $t = 3/5$) can be written as follows:

	x_3	$B - 459.4$	
x_4	11	1	$\frac{3}{5}$
x_1	$-10\frac{1}{11}$	$-10/11$	14
F	$-10\frac{1}{11}$	$-10/11^*$	0
x_2	$-10/11$	$-1/11$	$5\frac{3}{5}$

It looks as if x_2 were already at its minimum value, and this is indeed true as long as $B - 459.4$ remains at zero. But it is precisely this amount which we want to reduce again. F is always zero, and therefore we make it nonbasic in exchange for $B - 459.4$. Admittedly, the pivot is now negative, but the rules of transformation are not thereby changed. After the transformation, we introduce two new variables,

$$B - 459.4 + t_2 \quad \text{and} \quad t_2,$$

in the same way as t_1 was introduced with H_1 before, and this leads to

	x_3	F	t_2	
x_4	$-1/10$	$11/10$	0	$\frac{3}{5}$
x_1	0	-1	0	14
H_2	$111/10^*$	$-11/10$	-1	0
x_2	$1/10$	$-1/10$	0	$5\frac{3}{5}$

and, finally, to

	H_2	F	t_2	
x_4	$1/111$	$110/111$	$-1/111$	$\frac{3}{5}$
x_1	0	-1	0	14
x_3	$10/111$	$-11/111$	$-10/111$	0
x_2	$-1/111$	$-10/111$	$1/111$	$5\frac{3}{5}$

This means that $x_2 + t_2/111 = 5\frac{3}{5}$, and to make x_2 equal to 5 we must make $t_2 = 44.4$. This gives

$$x_2 = 5, \quad x_1 = 14, \quad B = 459.4 - 44.4 = 415,$$
$$x_4 = 1, \quad x_3 = 4.$$

The values are now all integral, but it is not certain that we have obtained the highest value of B under the circumstances. We must, therefore, also find the value of B such that the maximum of x_2 is 6, while x_1 remains at 14 (actually there is no such value of B), then change x_1 into other values and x_2 into compatible integer values, and so on. We shall not carry out the whole computation, but indicate its progress:

$$\text{Min } x_1 = 14, \qquad \text{Min } x_2 = 5, \qquad B = 415.$$
$$\text{Max } x_2 = 6, \qquad \text{not feasible.}$$
$$\text{Max } x_1 = 15, \qquad \text{Min } x_2 = 5, \qquad B = 405.$$
$$\text{Max } x_2 = 6, \qquad \text{not feasible.}$$

It is unnecessary to study $x_1 \geqslant 15$, because the corresponding values of B are already seen to be smaller than 415, a value that can be obtained. But we must investigate

$$\text{Min } x_1 = 13, \qquad \text{Min } x_2 = 5, \qquad B = 425.$$
$$= 12, \qquad\qquad\qquad\qquad = 435.$$
$$= 11, \qquad\qquad\qquad\qquad = 445.$$
$$= 10, \qquad\qquad\qquad\qquad = 455.$$
$$= 9, \qquad \text{Min } x_2 = 4, \qquad B = 354.$$

Hence the optimum is

$$x_1 = 10, \qquad x_2 = 5, \qquad B = 455.$$

We mention that Land and Doig use in their paper the Inverse Matrix method, but this does not alter the underlying ideas.

EXERCISES

10–1. Derive the constraints which make the region of Fig. 10–6 feasible.

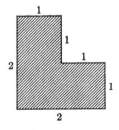

FIGURE 10–6

10–2. Consider Example 10–3 and (a) start by considering the variable x_5'; (b) consider x_1.

10–3. Consider those values of u and v in Exercise 9–2 for which

$$u + 2v \geqslant -1 \quad \text{and} \quad -4u + v \leqslant 4,$$

and find the maximum of the objective function

$$(2 + 2u + v)x_1 + (1 + u - v)x_2,$$

subject to the further condition that all variables have integer values.

10–4. Let a triangle 012 have sides of length

3 (1–2)
4 (0–2)
5 (0–1).

Lay pipes along its sides with capacities such that:

(i) the pipes leading out of 0 have a total capacity of at least 3;
(ii) the pipes leading out of 1 or of 2 have total capacities of either 2 or 3;
(iii) no pipe between any two vertices must have a capacity exceeding 2.

Only pipes of an integer number of capacity units are available, and the cost of a pipe is proportional to its capacity and to its length. It is required to determine the cheapest system of piping.

10–5. In Exercise 5–7 let the value of n be 15, and solve the problem for integer, non-negative values of the variables.

10–6. Solve the problem

$$3x_1 + 4x_2 + x_3 \leqslant 2,$$
$$x_1 + 3x_2 + 2x_3 \leqslant 1.$$
$$\text{Maximize} \quad 3x_1 + 6x_2 + 2x_3,$$

with x_1 and x_2 non-negative integers, and x_3 any non-negative value.

CHAPTER 11

STOCHASTIC LINEAR PROGRAMMING

Let us assume that it is known that the technological coefficients lie within given upper and lower limits, so that

$$a_{ij}^- \leqslant a_{ij} \leqslant a_{ij}^+,$$
$$b_j^- \leqslant b_j \leqslant b_j^+,$$
$$c_i^- \leqslant c_i \leqslant c_i^+,$$

where those values written with a minus or plus sign as superscript are known. It is then natural to ask what can be inferred as to the range of possible variation of the optimum of the objective function.

11–1 Range of values. In order to fix our ideas, let it be required to minimize an objective function

$$F(x) = c_1 x_1 + \ldots + c_n x_n,$$

subject to

$$\sum_i a_{ij} x_i \geqslant b_j \quad (j = 1, \ldots, m), \qquad x_i \geqslant 0.$$

It is our intention to find values $F^-(x)$ and $F^+(x)$ such that whatever the value of the coefficients a_{ij}, b_j, or c_i, within their given ranges, the resulting minimum of $F(x)$ satisfies the relationships

$$F^-(x) \leqslant F(x) \leqslant F^+(x).$$

Our presentation is based on a communication from J. V. Talacko and R. T. Rockafellar.

Consider all sets $x_1, \ldots, x_n$ satisfying the inequalities

$$\sum_i a_{ij}^- x_i \geqslant b_j^+ \quad (j = 1, \ldots, m),$$
$$x_i \geqslant 0 \quad (i = 1, \ldots, n). \tag{i}$$

All these sets will also satisfy the inequalities

$$\sum_i a_{ij} x_i \geqslant b_j, \tag{ii}$$

where the a_{ij} are any values within their given ranges, because (ii) can be written

$$\sum_i a_{ij}^- x_i - b_j^+ \geqslant (b_j^- - b_j^+) + \sum_i (a_{ij}^- - a_{ij}) x_i,$$

and, since the right-hand side is not positive, this is certainly true when the left-hand side is non-negative, as it is by virtue of (i).

Similarly, if $x_1, \ldots, x_n$ satisfy the inequalities (ii) for any given a_{ij} and b_j, then they satisfy *a fortiori*

$$\sum_i a_{ij}^+ x_i \geqslant b_j^- \qquad (j = 1, \ldots, m), \qquad \text{(iii)}$$

when $x_i \geqslant 0$.

EXAMPLE 11-1. Given

(i)   (iii)

$-x_1 + 2x_2 \geqslant 3, \qquad 2x_1 + 3x_2 \geqslant 2, \qquad \tfrac{1}{2}x_1 + 2\tfrac{1}{2}x_2 \geqslant 2\tfrac{1}{2},$

$x_1 + x_2 \geqslant 4, \qquad 3x_1 + 12x_2 \geqslant 2, \qquad 2x_1 + 6\tfrac{1}{2}x_2 \geqslant 3,$

$n = m = 2, \qquad -1 \leqslant a_{11} \leqslant 2, \qquad 1 \leqslant a_{12} \leqslant 3,$

$2 \leqslant a_{21} \leqslant 3, \qquad 1 \leqslant a_{22} \leqslant 12,$

$2 \leqslant b_1 \leqslant 3, \qquad 2 \leqslant b_2 \leqslant 4.$

The regions of points satisfying (i) and (iii) respectively are shown in the diagrams of Fig. 11-1, and so is a region corresponding to values of

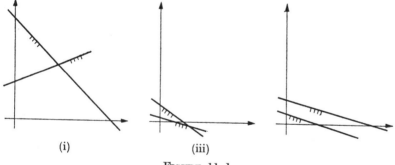

(i)   (iii)

FIGURE 11-1

the coefficients just half way between their upper and lower limits. The feasible region of the latter contains that of (i), and is contained in that of (iii).

Consider now the following values:

F_1 defined as the minimum of $c_1^- x_1 + \ldots + c_n^- x_n$, subject to (iii).

F_2 the same, but subject to any of (ii).

F_3 defined as the minimum of some $c_1 x_1 + \ldots + c_n x_n$, subject to any of (ii).

F_4 defined as the minimum of $c_1^+ x_1 + \ldots + c_n^+ x_n$, subject to any of (ii).

F_5 the same, but subject to (i).

F_1 cannot be larger than F_2 because the $x_1, \ldots, x_n$ satisfying (ii) satisfy also (iii).

F_2 cannot be larger than F_3 because for any non-negative $x_1, \ldots, x_n$ we have $c_1^- x_1 + \ldots + c_n^- x_n \leqslant c_1 x_1 + \ldots + c_n x_n$.

F_3 cannot be larger than F_4 for a similar reason, and

F_4 cannot be larger than F_5 because the $x_1, \ldots, x_n$ satisfying (i) satisfy also (ii).

We have thus proved that

$$F_1 \leqslant F_3 \leqslant F_5,$$

that is

the minimum of $c_1 x_1 + \ldots + c_n x_n$ subject to (ii)

is not smaller than

the minimum of $c_1^- x_1 + \ldots + c_n^- x_n$ subject to $\sum_i a_{ij}^+ x_i \geqslant b_j^-$,

and not larger than

the minimum of $c_1^+ x_1 + \ldots + c_n^+ x_n$ subject to $\sum_i a_{ij}^- x_i \geqslant b^+$.

Similarly, it can be shown that

the maximum of $\sum_i c_i x_i$ subject to $\sum_i a_{ij} x_i \leqslant b_j$

cannot be larger than

the maximum of $\sum_i c_i^+ x_i$ subject to $\sum_i a_{ij}^- x_i \leqslant b_j^+$,

nor smaller than

the maximum of $\sum_i c_i^- x_i$ subject to $\sum_i a_{ij}^+ x_i \leqslant b_j^-$.

In an abbreviated form, easily understood, we can write these relations as follows:

$$
\begin{matrix}
\text{Min } \sum_i c_i^- x_i \\
\sum_i a_{ij}^+ x_i \geqslant b_j^-
\end{matrix}
\leqslant
\begin{matrix}
\text{Min } \sum_i c_i x_i \\
\sum_i a_{ij} x_i \geqslant b_j
\end{matrix}
\leqslant
\begin{matrix}
\text{Min } \sum_i c_i^+ x_i \\
\sum_i a_{ij}^- x_i \geqslant b_j^+
\end{matrix},
$$

and

$$
\begin{matrix}
\text{Max } \sum_i c_i^- x_i \\
\sum_i a_{ij}^+ x_i \leqslant b_j^-
\end{matrix}
\leqslant
\begin{matrix}
\text{Max } \sum_i c_i x_i \\
\sum_i a_{ij} x_i \leqslant b_j
\end{matrix}
\leqslant
\begin{matrix}
\text{Max } \sum_i c_i^+ x_i \\
\sum_i a_{ij}^- x_i \leqslant b_j^+
\end{matrix}.
$$

11-2 Distribution problems and expected value problems. We shall present theorems which refer to constraints in the form of either equations, or inequalities. In order to deal with both forms simultaneously, we write the constraints as

$$\sum_{i=1}^{n} a_{ij}x_i + \sum_{k=1}^{p} d_{kj}y_k = b_j \qquad (j = 1, \ldots, m),$$

and the objective function as

$$\sum_i c_i x_i + \sum_k f_k y_k.$$

This subsumes equations (all $d_{kj} = 0$) and inequalities ($d_{kj} = 1$ or $= -1$ if $k = j$, and $= 0$ otherwise), but also more general cases.

If we assume that the technological coefficients are random variables, then we speak of *Stochastic Linear Programming*. In these circumstances the optimum of an objective function is also a random variable. It is then natural to try to determine its distribution, a task which is complicated by the fact that the set of basic variables producing the optimum is, in general, dependent on the values of the coefficients.

We shall call problems concerned with the distribution of the optimal value of the objective function, or with finding its parameters, *Distribution Problems*. They are also considered in papers by G. Tintner [116], M. M. Babbar [20], and H. M. Wagner [123]. A. Madansky, in [101], speaks here of "Wait-and-See" situations: We wait for an observation on the random variables to be available and then solve the nonstochastic linear programming problem based on these observed values.

Other practical applications have led to problems of a different type, which we call *Expected Value Problems*. They may be introduced by an argument explicitly stated by E. M. L. Beale [24] and also implicit in work by G. B. Dantzig [43] and by A. R. Ferguson and G. B. Dantzig [63].

Assume the constraints to be equations. If the coefficients are random variables, then it cannot be reasonably expected that the equations be precisely satisfied. But if any discrepancy between the two sides of any equation leads to a loss, then we can incorporate this into the objective function, and it is then the expected value of the latter which we shall want to minimize.

The resulting problem can also be regarded as a special case of non-stochastic linear programming, provided the random variables have discrete distributions. For if there is a known probability for any set of coefficients, then by introducing new variables for the discrepancies we can write down a scheme including all the constraints with the several possible sets of coefficients, and minimize the expected value of an objective function including fines for discrepancies. However, such a scheme would soon exceed the capacity of any available computing installation.

Dantzig, in [43], describes our Expected Value Problems as two-stage problems. We decide first on some values for the x_i, then observe the coefficients, and compensate for discrepancies between the two sides of the constraints. Hence Madansky's expression: "Here-and-Now" situation. As an example, consider the case dealt with by Ferguson and Dantzig. If the aircraft scheduled to fly on a given route is not sufficient to carry all the passengers who require transport, then a loss of fares is suffered by the company. On the other hand, over-supply of aircraft produces a loss which is different from that due to insufficient supply. Madansky mentions the following formulation of this type: Among all x_i and y_j whose probability of feasibility is at least p, find those y_j which minimize an objective function, and then determine the values of x_i which maximize the probability that the minimum of the objective function does not exceed a given value k.

We now prove two theorems contained in Beale's paper [24].

THEOREM 11-1. Let $x_1, \ldots, x_n$ be fixed, say $x_i = x_i^0 \geqslant 0$, and minimize

$$\sum_i c_i x_i^0 + \sum_k f_k y_k = C(x^0, y), \text{ say,}$$

subject to

$$\sum_{k=1}^p d_{kj} y_k = b_j - \sum_{i=1}^n a_{ij} x_i^0 \quad (j = 1, \ldots, m),$$

$$y_k \geqslant 0.$$

Then the minimum with regard to y_k of $C(x^0, y)$ is a convex function of $x_1, \ldots, x_n$.

In other words, in view of the definition of convexity, we want to show that

$$\min_{y_k} C(\lambda x' + (1 - \lambda)x'', y) \leqslant \lambda \min_{y_k} C(x', y) + (1 - \lambda) \min_{y_k} C(x'', y),$$

where $0 \leqslant \lambda \leqslant 1$. The minima are to be taken over all y_k which satisfy the constraints with $\lambda x_i' + (1 - \lambda)x_i''$, with x_i', and with x_i'', respectively.

Proof. Let the two minima on the right-hand side be obtained for y_k' and y_k'' respectively. Then

$$\sum_{k=1}^p d_{kj} y_k' = b_j - \sum_{i=1}^n a_{ij} x_i',$$

and

$$\sum_{k=1}^{p} d_{kj} y_k'' = b_j - \sum_{i=1}^{n} a_{ij} x_i'',$$

and hence also

$$\sum_{k=1}^{p} d_{kj}(\lambda y_k' + (1 - \lambda)y_k'') = b_j - \sum_{i=1}^{p} a_{ij}(\lambda x_i' + (1 - \lambda)x'').$$

Moreover,

$$C[\lambda x' + (1 - \lambda)x'', \lambda y' + (1 - \lambda)y'']$$
$$= \lambda C(x', y') + (1 - \lambda)C(x'', y'')$$
$$= \lambda \min_{y_k} C(x', y) + (1 - \lambda) \min_{y_k} C(x'', y).$$

Now $\lambda y_k' + (1 - \lambda)y_k''$ satisfy the constraints with $\lambda x_i' + (1 - \lambda)x_i''$, and hence the minimum of $C[\lambda x' + (1 - \lambda)x'', y]$ over all admissible y_k cannot be larger than the right-hand side. Q.E.D.

THEOREM 11–2. Let the assumptions be the same as in Theorem 11–1. Then the minimum with regard to y_k of $C(x^0, y)$ is a convex function of the coefficients a_{ij} and b_j.

In other words, we want to show that

$$\min_{y_k}{}_0 C(x^0, y) \leqslant \lambda \min_{y_k}{}_1 C(x^0, y) + (1 - \lambda) \min_{y_k}{}_2 C(x^0, y),$$

where $\min_s$ is to be computed under the constraints

$$\sum_{k=1}^{p} d_{kj} y_k = b_{js} - \sum_{i=1}^{n} a_{ijs} x_i^0 = B_{js}, \text{ say} \qquad (s = 0, 1, 2),$$
$$B_{j0} = \lambda B_{j1} + (1 - \lambda)B_{j2},$$
$$0 \leqslant \lambda \leqslant 1.$$

Proof. Let the two minima on the right-hand side be obtained by y_k' and y_k'' respectively. Then

$$\sum_{k=1}^{p} d_{kj} y_k' = B_{j1}, \qquad \sum_{k=1}^{p} d_{kj} y_k'' = B_{j2},$$

and hence also

$$\sum_{k=1}^{p} d_{kj}[\lambda y_k' + (1 - \lambda)y_k''] = \lambda B_{j1} + (1 - \lambda)B_{j2}.$$

Moreover

$$C[x^0, \lambda y' + (1 - \lambda)y''] = \lambda C(x^0, y') + (1 - \lambda)C(x^0, y'')$$
$$= \lambda \min_{y_k}{}_1 C(x^0, y) + (1 - \lambda) \min_{y_k}{}_2 C(x^0, y).$$

Now $\lambda y_k' + (1 - \lambda)y_k''$ satisfy the constraints with B_{j0}, and hence the minimum of $C(x^0, y)$ with y_k subject to those constraints cannot be larger than the right-hand side. Q.E.D.

The result shows that the average of a number of minimum values of the objective function, derived for various values of the coefficients, cannot be lower than that minimum value, derived from the average values of the coefficients.

These two theorems can be specialized in a variety of ways, and we shall derive relations which are, in fact, such generalizations. However, we shall establish them more directly.

In what follows we shall assume that only the right-hand sides of the constraints, the b_j, are random variables, and that their mean values are known.

We consider now a typical Distribution Problem ([cf. 118]). Let the constraints be

$$\sum_{i=1}^{n} a_{ij}x_i = b_j \qquad (j = 1, \ldots, m),$$

$$x_i \geqslant 0 \qquad (i = 1, \ldots, n),$$

and minimize

$$c_1 x_1 + \ldots + c_n x_n.$$

Denote the solution by

$$x_1^0, \ldots, x_n^0,$$

and write

$$c_1 x_1^0 + \ldots + c_n x_n^0 = M_0.$$

Assume now that the right-hand sides are subject to discrete random variation, and that thus there arise the several problems, each characterized by a different value of t:

$$\sum_{i=1}^{n} a_{ij}x_i = b_j + b_{jt} \qquad (j = 1, \ldots, m),$$

$$x_i \geqslant 0 \qquad (i = 1, \ldots, n).$$

$$\text{Minimize } c_1 x_1 + \ldots + c_n x_n.$$

We denote the solution of such a problem by

$$x_{1t}^0, \ldots, x_{nt}^0,$$

and write

$$c_1 x_{1t}^0 + \ldots + c_n x_{nt}^0 = M_{0t}.$$

Let the probability of a set $b_{1t}, \ldots, b_{mt}$ for a specified value of t be p_t; let

$$Eb_{jt} = \sum_{t} p_t b_{jt} = 0 \qquad \text{for all } j,$$

and assume that no set $b_{1t}, \ldots, b_{mt}$ makes the constraints (including $x_i \geqslant 0$) contradictory.

It will now be shown that it is generally true that

$$M_0 \leqslant \sum_t p_t M_{0t}.$$

Consider the dual of the original problem, viz.

$$\sum_{j=1}^{m} y_j a_{ij} \leqslant c_i \qquad (i = 1, \ldots, n).$$

$$\text{Maximize } \sum_j y_j b_j.$$

The solution set is denoted by $y_1^0, \ldots, y_m^0$.

Consider also a typical specimen of the dual of the modified problems. Thus

$$\sum_{j=1}^{m} y_j a_{ij} \leqslant c_i \qquad (i = 1, \ldots, n).$$

$$\text{Maximize } \sum_j y_j (b_j + b_{jt}).$$

Let its solution be

$$y_{1t}^0, \ldots, y_{mt}^0.$$

By the Duality theorem we have

$$\sum_j y_j^0 b_j = M_0,$$

and

$$\sum_j y_{jt}^0 (b_j + b_{jt}) = M_{0t} \qquad \text{for all } t.$$

All problems which we consider here have the same objective function. In the duals it is the constraints that are the same; all of them are satisfied by $y_1^0, \ldots, y_m^0$. This is the set that produces the maximum M_0, but not necessarily the maximum in a modified problem. Therefore

$$\sum_j y_j^0 (b_j + b_{jt}) \leqslant M_{0t} \qquad \text{for all } t.$$

By virtue of $Eb_{jt} = 0$ for all j, it follows that

$$M = \sum_j y_j^0 b_j = E \sum_j y_j^0 (b_j + b_{jt}) \leqslant E M_{0t}. \qquad \text{Q.E.D.}$$

This proof breaks down when either M_0 or any of the M_{0t} is infinite, because then no values y_j^0 or y_{jt}^0 exist. But if one of them is infinite, then all of them are and the relation is trivially true.

It is easily seen that the argument remains valid when the constraints are inequalities rather than equations. The result follows also from Theorem 11–2 if we put $x_i^0 = 0$ and interpret $\sum_k d_{ki} y_k$ as $\sum_i a_{ij} x_i$.

We turn now to an Expected Value problem. Find non-negative values $x_1, \ldots, x_n$ such that for all j and t simultaneously

$$\sum_{i=1}^{n} a_{ij}x_i \leqslant b_j + b_{jt},$$

where the mean of all the b_{jt} is zero for all j.
The constraints can also be written

$$\sum_{i=1}^{n} a_{ij}x_i \leqslant b_j + \min_t b_{jt},$$

and we assume that such values x_i exist. This is equivalent to looking for non-negative values

$$x_1, \ldots, x_n, \qquad x_{n+1,t}, \ldots, x_{n+m,t},$$

such that

$$\sum_{i=1}^{n} a_{ij}x_i + x_{n+j,t} = b_j + b_{jt} \quad \text{for all } t, \quad j = 1, \ldots, m.$$

The cost of a deficiency $x_{n+j,t}$ is assumed to be $f_j x_{n+j,t}$. We minimize the sum of the original objective function, and of the expected values of all positive fines $f_j x_{n+j,t}$, i.e.

$$\sum_{i=1}^{n} c_i x_i + \sum_t p_t \sum_{j=1}^{m} f_j x_{n+j,t} = \sum_i c_i x_i + \sum_t p_t \sum_j f_j (b_j + b_{jt} - \sum_i a_{ij}x_i),$$

where p_t is the probability of the values b_{jt} happening. Because the expected value of all the b_{jt} is zero for all j, the latter expression equals

$$\sum_i (c_i - \sum_j f_j a_{ij})x_i + \sum_j f_j b_j,$$

the last sum being a constant.
We are now faced with a nonstochastic linear programming problem. This simplification is due to our restricting possible deviations to one direction only. This may be justified by imagining that the cost attached to an excess of the left-hand over the right-hand side is so high that no such excess could possibly appear in the optimal solution.
Denote the solution set by $x_1', \ldots, x_n'$, and the minimum of the objective function by M_1.
It might occur to us that even if there are no random variations of the b_j, so that an equation could be precisely satisfied, it might be cheaper to allow deficiencies to occur and to accept the resulting fines. This leads to

$$\text{Minimize } \sum_i c_i x_i + \sum_j f_j (b_j - \sum_i a_{ij}x_i),$$

subject to

$$\sum a_{ij} x_i \leqslant b_j \qquad (j = 1, \ldots, m),$$

$$x_i \leqslant 0 \qquad (i = 1, \ldots, n).$$

Denote the optimal solution of this problem by $x_1'', \ldots, x_n''$, and the minimum by M_2. Then

$$M_2 \leqslant M_0 \qquad \text{and} \qquad M_2 \leqslant M_1,$$

as will now be shown.

The set x_i^0 satisfies the constraints of the last problem, and it makes the value of the objective function equal to M_0. Hence the latter can in any case be reached. But possibly some other set of variables gives an even lower value, and hence the first inequality. To compare M_1 and M_2, we notice that because the mean value of the b_{jt} is zero, we have

$$\min_t b_{jt} \leqslant 0,$$

so that

$$\sum_{i=1}^n a_{ij} x_i \leqslant b_j + \min_t b_{jt} \qquad (j = 1, \ldots, m)$$

implies the constraints

$$\sum_{i=1}^n a_{ij} x_i \leqslant b_j.$$

Hence the solution of the first constraints is also one of the second set of constraints, and M_2 will be at least as small as M_1.

The relationships which we have derived imply upper and lower bounds on some of the minima. Other bounds have been derived by A. Madansky in [100].

If nothing is known about the distribution of the b_{jt}, not even their means, then a minimax approach is indicated. Let again

$$\sum_{i=1}^n a_{ij} x_i \leqslant b_j + b_{jt} \qquad (j = 1, \ldots, m),$$

$$x_i \geqslant 0 \qquad (i = 1, \ldots, n),$$

and assume that these constraints can be satisfied for all t simultaneously. We want to find that set of variables x_i which minimizes the maximum cost, including fines, that could arise from any of the b_{jt}. Formally, we want to find x_i leading to

$$\min_x \max_t \left[\sum_{i=1}^n c_i x_i + \sum_{j=1}^m f_j \left(b_j + b_{jt} - \sum_{i=1}^n a_{ij} x_i \right) \right].$$

(Incidentally, it is well known from the Theory of Games that this $\min_x \max_t$ is never smaller than the corresponding $\max_t \min_x$.)

The search for variables x_i as defined is again a linear programming problem. This is seen as follows: Denote

$$\max_t \left[\sum_{i=1}^{n} c_i x_i + \sum_{j=1}^{m} f_j \left(b_j + b_{jt} - \sum_{i=1}^{n} a_{ij} x_i \right) \right]$$

by V. We then look for values of x_i which minimize V, subject to

$$\sum_{i=1}^{n} a_{ij} x_i \leqslant b_j + b_{jt} \qquad (j = 1, \ldots, m),$$

$$x_i \geqslant 0 \qquad (i = 1, \ldots, n),$$

$$\sum_{i=1}^{n} c_i x_i + \sum_{j=1}^{m} f_j (b_j + b_{jt} - \sum_{i=1}^{n} a_{ij} x_i) \leqslant V \quad \text{for all } t.$$

Write the last set of constraints

$$\sum_{i=1}^{n} c_i x_i + \sum_{j=1}^{m} f_j (b_j + b_{jt} - \sum_{i=1}^{n} a_{ij} x_i) + d_t = V,$$

$$d_t \geqslant 0 \qquad \text{(all } t\text{)}.$$

Take one particular value of t, say s, and subtract, in the last set, the constraints for s from all others, to obtain finally the following problem:

$$\sum_{i=1}^{n} a_{ij} x_i \leqslant b_j + b_{jt} \qquad (j = 1, \ldots, m),$$

$$x_i \geqslant 0 \qquad (i = 1, \ldots, n),$$

$$\sum_{j=1}^{m} f_j (b_{jt} - b_{js}) + d_t - d_s = 0 \quad \text{for all } t,$$

$$d_t \geqslant 0 \quad \text{for all } t \text{ (including } s\text{)}.$$

$$\text{Minimize } \sum_{i=1}^{n} (c_i - \sum_{j=1}^{m} f_j a_{ij}) x_i + \sum_{j=1}^{m} f_j (b_j + b_{js}) + d_s.$$

EXERCISES

11–1. Use the following example to illustrate $M_0 \leqslant \sum_t p_t M_{0t}$:

$$2x_1 + x_2 - x_3 = 1 + b_{1t},$$

$$x_1 + 2x_2 - x_3 = 1 + b_{2t},$$

$$\text{Minimize } x_1 + x_2 + x_3,$$

where $b_{1t} = -2, 0, \text{ or } 2, \; b_{2t} = -2, 0, \text{ or } 2$.

11-2. Use the following example to illustrate the relationship between M_0, M_1, and M_2:

$$2x_1 + x_2 - x_3 \leqslant 1 + b_{1t},$$
$$x_1 + 2x_2 - x_3 \leqslant 1 + b_{2t}.$$

Minimize $x_1 + x_2 + x_3 + Ef_1(1 + b_{1t} - 2x_1 - x_2 + x_3)$
$$+ Ef_2(1 + b_{2t} - x_1 - 2x_2 + x_3).$$

Take $f_1 = f_2 = 1$.

CHAPTER 12

NONLINEAR PROGRAMMING

It is natural, after the establishment of the methods of linear programming, to consider problems which do not fit into the framework of this technique, because either the constraints, or the objective function, cannot realistically be formulated in linear terms.

12–1 Definitions. We start with a few definitions. A function $f(x_1, \ldots, x_N)$, is called *convex* if for any pair of sets $(x_1', \ldots, x_N')$ and $(x_1'', \ldots, x_N'')$, and any t such that $0 < t < 1$, we have

$$f[tx_1' + (1 - t)x_1'', \ldots, tx_N' + (1 - t)x_N'']$$
$$\leqslant tf(x_1', \ldots, x_N') + (1 - t)f(x_1'', \ldots, x_N'').$$

A function is *strictly convex* if, for those values of t, the $\leqslant$ sign can always be replaced by a $<$ sign. A function is called *concave*, or *strictly concave*, respectively, if the inequality sign $\geqslant$ or $>$ holds. The negative of a convex function is a concave function, and hence the problem of minimizing a convex function is equivalent to that of maximizing a concave one. A linear function is concave (though not strictly), and also convex (not strictly). A sum of convex functions is also convex, and similarly a sum of concave functions is concave.

If $f(x) = f(x_1, \ldots, x_n)$ is convex, then for $0 < t \leqslant 1$

$$f(x') \geqslant f(x'') + \frac{1}{t} \left\{ f[x'' + t(x' - x'')] - f(x'') \right\},$$

and, therefore, as t approaches zero

$$f(x') \geqslant f(x'') + \sum_i \frac{\partial f}{\partial x_i} (x'')(x_i' - x_i''). \tag{12–1}$$

Similarly, if $f(x)$ is concave, the inequality sign is reversed:

$$f(x') \leqslant f(x'') + \sum_i \frac{\partial f}{\partial x_i} (x'')(x_i' - x_i''). \tag{12–2}$$

It is easily seen that the problem of maximizing a convex function, or that of minimizing a concave one, offers difficulties. (The diagrams of Fig. 12–1 refer to the case $N = 1$.) For instance, if we want to find the lowest point of a convex curve within a given range, then this lowest point lies in the direction of lower values, wherever we start from. But

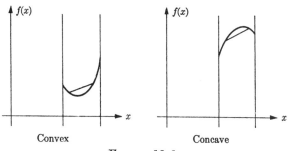

Convex   Concave

FIGURE 12–1

the lowest point of a concave curve within a given range will lie at one
of the ends of the range, and we do not always know in which direction
to move from any starting point, in order to get at the right one.

12–2 Approximations. We investigate first methods for approxi-
mating the solutions of nonlinear problems by those of linear ones.
To fix our ideas, we shall make our remarks in terms of minimizing
problems.

Let the objective function be a sum of terms $\phi_i(x_i)$, each of them
dependent on one single variable (cf. [37]). We approximate each such
term by a succession of linear expressions, each of them valid within a
given interval (Fig. 12–2). If the points at which the straight-line

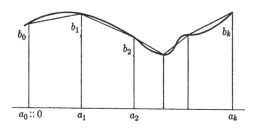

FIGURE 12–2

portions join have coordinates a_i, b_i then the successive straight lines
have equations

$$y = b_i + \frac{b_{i+1} - b_i}{a_{i+1} - a_i}(x - a_i),$$

or

$$y = b_i + s_{i+1}(x - a_i), \text{ say,}$$

valid for
$$a_i \leqslant x \leqslant a_{i+1} \qquad (i = 0, 1, \ldots, k - 1).$$

This can be written
$$x = x_1 + \ldots + x_k,$$
$$y = b_0 + s_1 x_1 + \ldots + s_k x_k,$$

with the proviso that $0 \leqslant x_i \leqslant a_i - a_{i-1}$ and that if
$$x_i < a_i - a_{i-1},$$
then
$$x_{i+1} = 0.$$

If $s_1 \leqslant s_2 \leqslant \ldots \leqslant s_k$, then this proviso may be omitted, because no x_i will appear in the minimum with a positive value unless all previous x_i have reached their upper limit. But if the s_i are not monotonically increasing, then we must state the proviso explicitly. It is equivalent to
$$0 \leqslant x_i \leqslant a_i - a_{i-1},$$
and either
$$x_i \geqslant a_i - a_{i-1} \qquad \text{(i.e. the equal sign holds)}$$
or
$$x_{i+1} = 0.$$

Using the ideas of discrete linear programming (see Chapter 10), the either-or condition can be written
$$x_i - d_i(a_i - a_{i-1}) \geqslant 0,$$
$$-x_{i+1} + d_i(a_{i+1} - a_i) \geqslant 0,$$
$$d_i = 0 \text{ or } 1 \qquad \text{for } i = 1, \ldots, k - 1.$$

A slightly different formulation is contained in [104]:
$$d_i(a_i - a_{i-1}) - x_i \geqslant 0,$$
$$x_i - d_{i+1}(a_i - a_{i+1}) \geqslant 0 \qquad \text{for } i = 1, \ldots, k - 1.$$

This has the same effect, because if $x_i < a_i - a_{i-1}$, then $d_{i+1} < 1$, and hence $= 0$. Consequently $-x_{i+1} \geqslant 0$, and hence $= 0$.

Yet another formulation was proposed by G. B. Dantzig as follows:
$$t_0 + t_1 + \ldots + t_k = 1, \qquad t_i \geqslant 0 \quad (i = 0, 1, \ldots, k),$$
$$x = t_0 a_0 + \ldots + t_k a_k,$$
$$y = t_0 b_0 + \ldots + t_k b_k,$$

where only one pair t_i and t_{i+1} may be positive for a single i. This condition may be written

$$
\begin{aligned}
d_0 + d_1 + \quad\quad\quad + d_{k-1} &= 1, \\
d_0 \quad\quad\quad\quad\quad &\geqslant t_0, \\
d_0 + d_1 \quad\quad\quad &\geqslant t_1, \\
d_1 + d_2 \quad\quad &\geqslant t_2,
\end{aligned}
$$

.
.
.

$$
\begin{aligned}
d_{k-2} + d_{k-1} &\geqslant t_{k-1}, \\
d_{k-1} &\geqslant t_k,
\end{aligned}
$$

all d_i equal to 0 or to 1.

These conditions exclude the possibility of two not successive a_i entering into the definition of x (cf. [53]).

H. O. Hartley, in [81], deals with problems where the constraints are nonlinear. An example will make the procedure clear.

We consider here functions of three variables, subject to constraints

$$ h_{1t}(x_1) + h_{2t}(x_2) + h_{3t}(x_3) \leqslant c_t \quad\quad (t = 1, \ldots, m). $$

It is also assumed that all $h_{it}(x_i)$ are convex functions of their variables. Approximate such a function by a broken line, the sth portion of which is defined by

$$ {}_s a_{it} + {}_s b_{it} x_i \quad\quad (s = 1, \ldots, m). $$

We describe here the procedure for $m = 2$. Write

$$ h_{1t}(x_1) \leqslant c_1 - h_{2t}(x_2) - h_{3t}(x_3). $$

Because all extensions of the segments are below their curve we have

$$
\begin{aligned}
{}_1 a_{1t} + {}_1 b_{1t} x_1 &\leqslant c_t - h_{2t}(x_2) - h_{3t}(x_3), \\
{}_2 a_{1t} + {}_2 b_{1t} x_1 &\leqslant c_t - h_{2t}(x_2) - h_{3t}(x_3).
\end{aligned}
$$

Now transfer $h_{2t}(x_2)$ to the left-hand side and deal with it in the same way, so that from the first inequality we have

$$
\begin{aligned}
{}_1 a_{2t} + {}_1 b_{2t} x_2 &\leqslant c_t - {}_1 a_{1t} - {}_1 b_{1t} - h_{3t}(x_3), \\
{}_2 a_{2t} + {}_2 b_{2t} x_2 &\leqslant c_t - {}_1 a_{1t} - {}_1 b_{1t} - h_{3t}(x_3),
\end{aligned}
$$

and from the second

$$
\begin{aligned}
{}_1 a_{2t} + {}_1 b_{2t} x_2 &\leqslant c_t - {}_2 a_{1t} - {}_2 b_{1t} - h_{3t}(x_3), \\
{}_2 a_{2t} + {}_2 b_{2t} x_2 &\leqslant c_t - {}_2 a_{1t} - {}_2 b_{1t} - h_{3t}(x_3).
\end{aligned}
$$

Treating $h_{3t}(x_3)$ in the same way, we obtain finally eight inequalities, of which the first and the last are

$$_1a_{3t} + {}_1b_{3t}x_3 \leqslant c_t - {}_1a_{1t} - {}_1a_{2t} - {}_1b_{1t}x_1 - {}_1b_{2t}x_2$$

$$\cdot$$
$$\cdot$$
$$\cdot$$

$$_2a_{3t} + {}_2b_{3t}x_3 \leqslant c_t - {}_2a_{1t} - {}_2a_{2t} - {}_2b_{1t}x_1 - {}_2b_{2t}x_2.$$

In general, one obtains m^n inequalities for n variables. They are linear, but clearly further processing is necessary to deal with their enormous numbers.

12–3 Quadratic programming. We turn now to *quadratic* programming. This term will be applied to mean the minimization of a convex, or the maximization of a concave, quadratic expression, while the constraints are still linear in all cases we deal with.

A quadratic expression in $x_1, \ldots, x_n$ can be convex, or concave, or neither. The diagrams of Fig. 12–3 illustrate such cases.

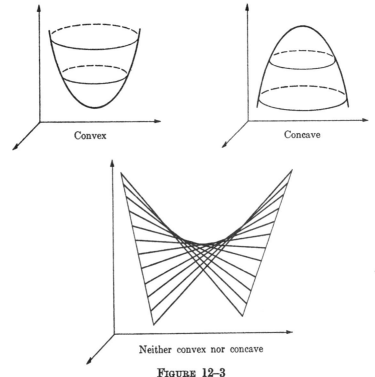

Convex

Concave

Neither convex nor concave

FIGURE 12–3

It is known that a *quadratic form*, i.e. a function with only terms of second degree in all variables, can be written as a sum or difference of squares of independent homogeneous linear expressions. For instance

$$5x_1^2 - 16x_1x_2 + 13x_2^2 = (x_1 - 2x_2)^2 + (2x_1 - 3x_2)^2.$$

It is evident that if a form in n variables can be transformed into a sum of n such squares with positive coefficients, then the function can only be zero if all squares are, and hence only if all $x_i = 0$. Otherwise it is positive. Such a function is called *positive definite*. If it can be transformed into a sum of less than n such squares, all with positive coefficients, then it cannot have negative values, but it can be zero even when not all variables vanish. It is then *positive semidefinite*.

Similarly, a quadratic form is *negative definite* or *semidefinite* if it can be transformed into a sum of n, or of less, squares with negative coefficients.

A positive semidefinite quadratic form is convex, and a positive definite form is strictly convex. This will now be shown. Let the function be

$$f(x_1, \ldots, x_n) = \sum_i \sum_j a_{ij} x_i x_j,$$

and let $a_{ij} = a_{ji}$ (this is not a restriction of generality). Compute

$$f[tx_1' + (1 - t)x_1'', \ldots, tx_n' + (1 - t)x_n''] =$$
$$t^2 \sum_i \sum_j a_{ij} x_i' x_j' + 2t(1 - t) \sum_i \sum_j a_{ij} x_i' x_j'' + (1 - t)^2 \sum_i \sum_j a_{ij} x_i'' x_j''.$$

Subtract this from

$$tf(x_1', \ldots, x_n') + (1 - t)f(x_1'', \ldots, x_n'') = t\sum_i \sum_j a_{ij} x_i' x_j' + (1 - t)\sum_i \sum_j a_{ij} x_i'' x_j''.$$

The difference reduces to

$$t(1 - t)\sum_i \sum_j a_{ij}(x_i' - x_i'')(x_j' - x_j'').$$

Hence, if we are dealing with a positive definite form, then it is strictly convex and if the form is positive semidefinite, then it is convex. A similar theorem holds for negative definite forms and for concavity.

We shall deal with a minimizing problem, and write the objective function as follows:

$$\begin{aligned} C = c_0 &+ c_1 x_1 + \ldots + c_n x_n \\ &+ (c_1 + c_{11} x_1 + \ldots + c_{1n} x_n)x_1 \\ &\quad \cdot \\ &\quad \cdot \\ &\quad \cdot \\ &+ (c_n + c_{n1} x_1 + \ldots + c_{nn} x_n)x_n, \end{aligned}$$

where $c_{ij} = c_{ji}$. This can also be written

$$c_0 + (c_1 x_1 + \ldots + c_n x_n) + \frac{1}{2}\left(x_1 \frac{\partial C}{\partial x_1} + \ldots + x_n \frac{\partial C}{\partial x_n}\right).$$

The process which we explain first (see [24] and, in more detail, [28]) has many similarities with the Simplex method. The procedure of finding a first feasible solution is the same, because it depends only on the constraints. In order to see whether the increase of a nonbasic variable would be useful, we observe, in the Simplex method, the constant coefficients of such variables in the expression of the objective function. These coefficients are, in the linear case, the partial derivatives of the objective function. Similar use is being made of the partial derivatives in quadratic programming, but the latter are then not constants any more.

Let C be expressed in terms of the nonbasic variables (with zero values):

$$
\begin{aligned}
C = {} & z_{00} + z_{0m+1} x_{m+1} + \ldots + z_{0n} x_n \\
& + (z_{0m+1} + z_{m+1,m+1} x_{m+1} + \ldots + z_{m+1n} x_n) x_{m+1} \\
& \quad\cdot \\
& \quad\cdot \\
& \quad\cdot \\
& + (z_{0n} + z_{n,m+1} x_{m+1} + \ldots + z_{nn} x_n) x_n.
\end{aligned}
$$

We want to minimize it.

A nonbasic variable can be usefully increased if the partial derivative with regard to it is negative. In the point just reached, the value of this derivative, with regard to x_{m+j}, equals $2z_{0,m+j}$. If this is negative, then an increase of x_{m+j} is useful, until either

$$u_{m+j} = z_{0m+j} + z_{m+j,m+1} x_{m+1} + \ldots + z_{m+j,n} x_n,$$

or one of the basic variables becomes zero. To watch both these points conveniently, we introduce u_{m+j} as a new variable. It can take positive, zero, or negative values and we call it therefore a *free* variable. We become interested in such a variable when it is the coefficient of a nonbasic variable which we want to make basic. We then make this new free variable nonbasic, and treat it like any other, except that sometimes its reduction rather than its increase might become useful. Once a hitherto nonbasic free variable has been made basic again, it can be ignored, because it does not matter if it becomes negative through a change in the value of another variable, nor are we concerned with its value in the final solution. We illustrate the procedure by an example in which $n = 2$.

EXAMPLE 12–1. Minimize

$$C = 10 + [2(x_1 - 5) - 4(x_2 - 3)]^2 + [2(x_1 - 5) + (x_2 - 3)]^2$$
$$= 183 - 44x_1 - 42x_2 + 8x_1^2 - 12x_1x_2 + 17x_2^2,$$

subject to

$$2x_1 + x_2 + x_3 = 10 \qquad \text{with } x_1, x_2, x_3 \geqslant 0.$$

The objective function can also be written

$$183 - 22x_1 - 21x_2 + (-22 + 8x_1 - 6x_2)x_1 + (-21 - 6x_1 + 17x_2)x_2.$$

If there were no constraints, then we could solve our problem by an application of the differential calculus:

$$\frac{1}{2}\frac{\partial C}{\partial x_1} = -22 + 8x_1 - 6x_2 = 0,$$

$$\frac{1}{2}\frac{\partial C}{\partial x_2} = -21 - 6x_1 + 17x_2 = 0,$$

i.e.

$$x_1 = 5, \qquad x_2 = 3.$$

We see also from the first expression of C given above that this point is, indeed, the centre of all elliptic contours of constant C, and that hence the minimum of C is reached in this point. The minimum value is 10.

The contour for $C = 183$ passes through the origin and, also, through the points

$$x_1 = 0, \qquad x_2 = 42/17,$$

and

$$x_1 = 11/2, \qquad x_2 = 0.$$

The rectangular axes of all contours are on the straight lines defined by

$$x_1 - 2x_2 + 1 = 0 \qquad \text{and} \qquad 2x_1 + x_2 - 13 = 0.$$

Because of the constraint, the point $x_1 = 5$, $x_2 = 3$ is not in the feasible region, and we must find the lowest point of C whose coordinates satisfy the constraints. We start in point A, the origin (Fig. 12–4). Looking at the partial derivatives in the expression for C, it seems useful to increase either x_1 or x_2, since the constants in both $\partial C/\partial x_1$

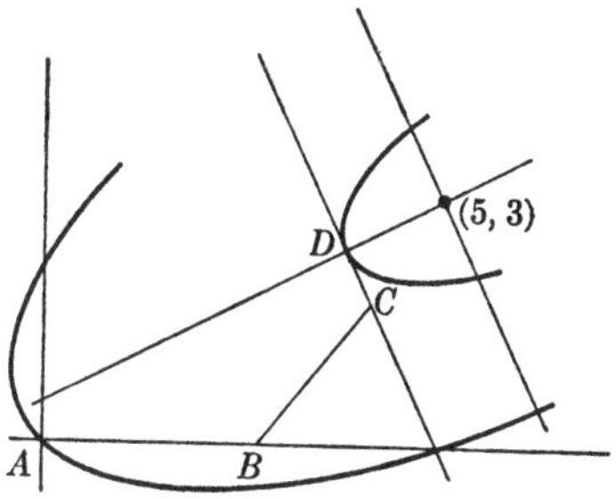

FIGURE 12–4

and $\partial C/\partial x_2$ are negative. We choose the former. Its increase must be stopped when either

$$-22 + 8x_1 - 6x_2 = 0$$

or

$$x_3 = 10 - 2x_1 - x_2 = 0.$$

Since we keep x_2 fixed at zero, the first equation holds for a smaller x_1. Therefore we introduce the new variable

$$u_1 = -22 + 8x_1 - 6x_2,$$

and x_1 will be increased until $u_1 = 0$, i.e. until $x_1 = 2.75$.

Notice that in the same way as we would have ignored a positive coefficient of x_1 in the expression of x_3, we would have been also justified in ignoring a negative coefficient of x_1 in the expression of a partial derivative. It would have indicated that there was no limit to an increase of x_1, as far as the surface C was concerned. This could only have happened if the quadratic form had been positive semidefinite, and its contours hence parallel lines rather than ellipses.

The fact that it was $u_1 = 0$ that stopped us from increasing x_1 further means that we have now reached a point beyond which the surface C would ascend. The straight line $u_1 = 0$ passes, of course, through the (not feasible) absolute minimum of C.

Because x_1 has become positive, and u_1 has become zero, we exchange these two variables. Thus

$$
\begin{aligned}
x_1 = &\quad 2.75 + 0.125u_1 + \ 0.75x_2, \\
x_3 = &\quad 4.5 \ - 0.25 \ u_1 - \ 2.5 \ x_2, \\
C = &\quad 122.5 \ - 2.75 \ u_1 - 37.5 \ x_2 \\
&+ (\qquad\qquad u_1 \qquad\quad)x_1 \\
&+ (-37.5 \ - 0.75 \ u_1 + 12.5 \ x_2)x_2 \\
= &\quad 122.5 \qquad\qquad - 37.5 \ x_2 \\
&+ (\qquad 0.125u_1 \qquad\quad)u_1 \\
&+ (-37.5 \qquad\quad + \ 12.5x_2)x_2.
\end{aligned}
$$

The value of x_2 is zero, and the point $(2.75, 0)$ is B in the diagram.

Now x_2 is worth increasing, up to the smaller of $4.5/2.5$ (when x_3 becomes zero) and $37.5/12.5$, when the partial derivative $-37.5 + 12.5x_2$

becomes zero. The smaller of these ratios is the first, so that we exchange x_2 and x_3. This gives

$$
\begin{aligned}
x_1 &= && 4.1 + 0.05\ u_1 - 0.3x_3, \\
x_2 &= && 1.8 - 0.1\ u_1 - 0.4x_3, \\
C &= && 55\ \ + 3.75\ u_1 + 15\ \ x_3 \\
&\ \ + (&& 0.125u_1 \quad\quad)u_1 \\
&\ \ + (-15 && -1.25\ u_1 - 5\ \ x_3)x_2 \\
&= && 28\ \ + 1.5\ \ u_1 + 6\ \ x_3 \\
&\ \ + (&& 1.5 + 0.25\ u_1 + 0.5x_3)u_1 \\
&\ \ + (&& 6\ \ + 0.5\ \ u_1 + 2\ \ x_3)x_3.
\end{aligned}
$$

The point $(4.1, 1.8)$ is marked C in the diagram of Fig. 12–4.

It is now necessary to decrease u_1 until either

$$4.1 + 0.05u_1 = 0 \qquad \text{or} \qquad 1.5 + 0.25u_1 = 0.$$

This happens the first time when $u = -6$, and then the new variable $u_2 = 1.5 + 0.25u_1 + 0.5x_3$ equals zero. [Of course, the line $u_2 = 0$ passes also through the absolute minimum at $(5, 3)$.] We exchange u_1 for u_2. Thus

$$
\begin{aligned}
u_1 &= && -6\ \ + 4\ \ u_2 - 2\ \ x_3 \quad \text{(This can now be ignored.)}, \\
x_1 &= && 3.8 + 0.2u_2 - 0.4x_3, \\
x_2 &= && 2.4 - 0.4u_2 - 0.2x_3, \\
C &= && 19\ \ + 6\ \ u_2 + 3\ \ x_3 \\
&\ \ + (&& u_2 \quad\quad)u_1 \\
&\ \ + (\ 3 && + 2\ \ u_2 + \quad x_3)x_3 \\
&= && 19 \quad\quad\quad + 3\ \ x_3 \\
&\ \ + (&& 4\ \ u_2 \quad\quad)u_2 \\
&\ \ + (\ 3 && + \quad x_3)x_3.
\end{aligned}
$$

Now an increase of x_3 or a change of u_2 would not reduce C, and we have therefore reached the optimum in point D with coordinates $x_1 = 3.8$ and $x_2 = 2.4$.

We add a few remarks concerning the computation.

We have, in Example 12–1, carried out the transformation of the objective function in two steps, first introducing the new variable only within the brackets, and then also substituting for the old variable outside one of the brackets. The second step can be viewed as an operation on columns, as distinct from the first, an operation on rows. The following general explanation will make this clear.

Let the objective function be written

$$C = a_{00} + a_{01}x_1 + \ldots + a_{0s}x_s$$
$$+ (a_{10} + a_{11}x_1 + \ldots + a_{1s}x_s)x_1$$
$$\cdot$$
$$\cdot$$
$$\cdot$$
$$+ (a_{s0} + a_{s1}x_1 + \ldots + a_{ss}x_s)x_s,$$

where the x_1 may be free or sign-restricted.

Let a new variable, y, be introduced by

$$y = b_0 + b_1x_1 + \ldots + b_sx_s,$$

in exchange for x_s. We solve the latter equation for x_s. Thus

$$x_s = c_0 + c_1x_1 + \ldots + c_{s-1}x_{s-1} + c_sy.$$

Substituting for x_s within the brackets, but not in the x_s outside the last bracket, we obtain

$$(a_{00} + \ldots + a_{0s}c_0) + (a_{01} + \ldots + a_{0s}c_1)x_1 + \ldots + a_{0s}c_sy$$
$$+ [(a_{10} + \ldots + a_{1s}c_0) + (a_{11} + \ldots + a_{1s}c_1)x_1 + \ldots + a_{1s}c_sy]x_1$$
$$\cdot$$
$$\cdot$$
$$\cdot$$
$$+ [(a_{s0} + \ldots + a_{ss}c_0) + (a_{s1} + \ldots + a_{ss}c_1)x_1 + \ldots + a_{ss}c_sy]x_s.$$

We transpose now the matrix of coefficients, i.e. we write

$$(a_{00} + \ldots + a_{0s}c_0) + (a_{10} + \ldots + a_{1s}c_0)x_1 + \ldots + (a_{s0} + \ldots + a_{ss}c_0)x_s$$
$$+ [(a_{01} + \ldots + a_{0s}c_1) + (a_{11} + \ldots + a_{1s}c_1)x_1 + \ldots + (a_{s1} + \ldots + a_{ss}c_1)x_s]x_1$$
$$\cdot$$
$$\cdot$$
$$\cdot$$
$$+ [a_{0s}c_s + a_{1s}c_sx_1 + \ldots + a_{ss}c_sx_s]y,$$

and substitute again for x_s. The resulting matrix of coefficients is for $s = 2$ (the essential features remain the same for any s):

$$\begin{pmatrix} a_{00}+a_{02}c_0+a_{20}c_0+a_{22}c_0c_0, & a_{10}+a_{12}c_0+a_{20}c_1+a_{22}c_0c_1, & a_{20}c_2+a_{22}c_0c_2 \\ a_{01}+a_{02}c_1+a_{21}c_0+a_{22}c_0c_1, & a_{11}+a_{12}c_1+a_{21}c_1+a_{22}c_1c_1, & a_{21}c_2+a_{22}c_1c_2 \\ a_{02}c_2+a_{22}c_0c_2, & a_{12}c_2+a_{22}c_1c_2, & a_{22}c_2c_2 \end{pmatrix}.$$

We see from this that if the original matrix of coefficients was symmetrical, i.e. $a_{ij} = a_{ji}$, then so is the new one. We see, also, that if the new variable y is free, say $\frac{1}{2}(\partial C/\partial x_s)$, i.e. if

$$y = a_{s0} + a_{s1}x_1 + \ldots + a_{ss}x_s,$$

and hence

$$x_s = \frac{-a_{s0}}{a_{ss}} - \frac{a_{s1}x_1}{a_{ss}} - \ldots + \frac{y}{a_{ss}},$$

then the new variable y will only appear in a term

$$a_{ss}c_s^2 y^2 = \frac{y^2}{a_{ss}}$$

in the last expression. In other words if

$$y = \frac{1}{2}\frac{\partial C}{\partial x_s},$$

then the other coefficients in the row and in the column of y in the final matrix of coefficients will be zero, while the entries in the remaining rows and columns will appear already after the first transformation (concerning the rows), and will not be altered by the second transformation, concerning the columns.

We have not yet proved that our procedure terminates after a finite number of steps. We can avoid degeneracy by some perturbation method, as in the Simplex routine, but we must still convince ourselves that the introduction of new free variables at strategic points does not continue providing ever new nonbasic variables. We shall prove that this does not happen, if a rule is obeyed concerning the choice of nonbasic variables to become basic. This rule demands that if there exist nonbasic free variables u_i, such that $\partial C/\partial u_i \neq 0$, then one of these must be made basic, either by increasing or by decreasing it, whichever reduces C. A new nonbasic free variable will thus be introduced:

$$\frac{\partial C}{\partial u_i} = u_j, \text{ say.}$$

An expression for the objective function in which the linear terms contain no free variable is said to be in *standard form*. To begin with, there exists no free variable at all. The following types of transformation can arise:

(i) introduction of a new nonbasic free variable, in exchange for a sign-restricted one (first step in the example)—the new free variable will not appear in a linear term, as has just been proved.

(ii) exchange of two sign-restricted variables (second step in the example)—a linear term in a free variable can appear, but then it will lead, by our rule, to

(iii) the introduction of a new free variable, not in a linear term, and to the disappearance of a free variable which had been in a linear term (third step in the example).

As long as there are linear terms in free variables [they were all introduced on the occasion of an exchange of type (ii)], steps of type (iii) will be made. Hence, after a finite number of steps, we reach again a standard form. But there can exist only a finite number of these because, in each of them, the set of sign-restricted nonbasic variables defines uniquely the value of C, and this value is being steadily reduced (perhaps using a perturbation method). Hence, after a finite number of steps, we reach the minimum.

A parametric method has been developed by H. S. Houthakker (see [88]) for quadratic programming which is applicable when all coefficients of the constraints are non-negative. He calls it the *capacity method*, and we give here a brief sketch of it, illustrating it by the same example which we have just solved.

EXAMPLE 12–2. Minimize

$$C = \quad 183 - 22x_1 - 21x_2$$
$$+ (-22 + 8x_1 - 6x_2)x_1$$
$$+ (-21 - 6x_1 + 17x_2)x_2,$$

subject to

$$2x_1 + x_2 \leqslant 10 \quad \text{for non-negative } x_1 \text{ and } x_2.$$

A parameter, t say, appears in a further constraint

$$x_1 + x_2 = t.$$

We start from $t = 0$, and increase it continuously. When t is large enough, then this constraint is redundant as an addition to other constraints, with non-negative coefficients. Also, if the minimum of C lies within the feasible region, then there will be a value of t such that $x_1 + x_2 = t$ is satisfied by the coordinates of that point. Thus the addition of the parametric constraint does not modify the problem in its essence.

The only feasible point when $t = 0$ is the origin. Consider, in this point, $\frac{1}{2}(\partial C/\partial x_1)$, and $\frac{1}{2}(\partial C/\partial x_2)$. The absolute value of the first is the larger one, and therefore we increase x_1 with t. Thus we proceed until either a constraint stops us, or it becomes advisable to increase x_2 as well. Therefore we investigate which happens first, either

$$x_2 = 0, \quad x_1 + x_2 = t, \quad 2x_1 + x_2 = 10$$

or

$$x_2 = 0, \quad x_1 + x_2 = t, \quad -22 + 8x_1 - 6x_2 = -21 - 6x_1 + 17x_2,$$

i.e. we compare

$$x_1 = t = 5 \quad \text{and} \quad x_1 = t = \tfrac{1}{14}.$$

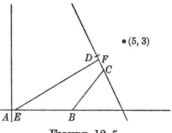

FIGURE 12–5

The value $1/14$ is the smaller one, and therefore for $t \leqslant 1/14$ we have $x_1 = t$, $x_2 = 0$ (point E in Fig. 12–5), but afterwards we proceed on the line

$$\frac{\partial C}{\partial x_1} = \frac{\partial C}{\partial x_2},$$

which carries us through the points

$$x_1 + x_2 = t, \qquad 14x_1 - 23x_2 = 1,$$

as long as the constraint

$$2x_1 + x_2 \leqslant 10$$

is satisfied. This is so until

$$t = 6.15, \qquad x_1 = 3.85, \qquad x_2 = 2.3 \text{ (Point } F).$$

From now on we remain on the line $2x_1 + x_2 = 10$, i.e. we maximize C subject to this equality constraint. Thus

$$\frac{\partial}{\partial x_1} C - \lambda(10 - 2x_1 - x_2) = 0,$$

$$\frac{\partial}{\partial x_2} C - \lambda(10 - 2x_1 - x_2) = 0,$$

$$x_1 + x_2 = t, \qquad 2x_1 + x_2 = 10.$$

The solution is

$$x_1 = 3.8, \qquad x_2 = 2.4, \qquad t = 6.2, \qquad \lambda = 6 \text{ (Point } D, \text{ as before).}$$

A comparison of Beale's and Houthakker's methods is contained in the following table:

	x_1	x_2	C		x_1	x_2	C
A	0	0	183	A	0	0	183
B	2.75	0	122.5	E	$\frac{1}{14}$	0	$179\frac{44}{49}$
C	4.1	1.8	28	F	3.85	2.3	19.25
D	3.8	2.4	19	D	3.8	2.4	19

12-4 Lagrangian methods. It is now our aim to exhibit a different approach to the optimization of a nonlinear function. By way of introduction, we consider a convex function of one single variable, $f(x)$. Let x be restricted to a finite portion of the axis of abscissae, and

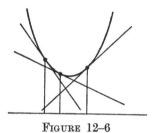

let the tangents to the curve $y = f(x)$ be given at a number of points (Fig. 12-6).

Imagine that verticals are drawn at feasible values of x. Find on each vertical the highest of its intersections with the given tangents, and find that vertical on which this highest intersection is lowest. This point is an approximation to the lowest point on the curve. If its vertical has abscissa x_0, then we draw a further tangent in $x_0, f(x_0)$ and repeat the process.

FIGURE 12-6

A similar method, due to E. W. Cheney and A. A. Goldstein, proceeds for one independent variable as follows (Fig. 12-7): Take the lowest point of intersection, which is here C. Find the highest line above it (AB) and the points on it in which it intersects the two lines through C. These are the points A and B. Delete the line through C and the lower

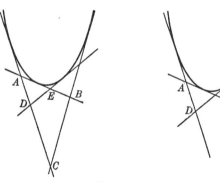

FIGURE 12-7

of those two points (i.e. B). Thus carry on. At the next step the lowest intersection in our example is D and the highest line above it is AE; we delete DE and are left with just one intersection, A. If we now draw a tangent through the point with the same abscissa as A, and repeat the procedure, then we approximate to the required minimum, provided certain simple conditions hold.

We turn now to a fundamental theorem of great theoretical interest, and which has also been used for purposes of numerical computation. To set the stage, we mention some results of the differential calculus.

Let a function $g(x_1, \ldots, x_n)$ be maximized, or minimized, under the side conditions

$$f_j(x_1, \ldots, x_n) = 0 \qquad (j = 1, \ldots, m).$$

We must then find the extremum of

$$g(x_1, \ldots, x_n) + \sum_j y_j f_j(x_1, \ldots, x_n),$$

which involves setting the partial derivatives of this function equal to zero. Further criteria allow us to decide whether any of the solutions is either a minimum, or a maximum, or neither. The y_j are called *Lagrange multipliers.*

When the variables are constrained by inequalities rather than by equations, or by both types, then the Lagrange method cannot be applied, but an appropriate extension of it has been developed by H. W. Kuhn and A. W. Tucker [98]. This will now be explained. Let

$$g(x) = g(x_1, \ldots, x_n)$$

and

$$f_j(x) = f_j(x_1, \ldots, x_n),$$

for $j = 1, \ldots, m$, be convex and differentiable functions, and let x_i be restricted to non-negative values.

THEOREM 12–1. $x^0 = (x_1^0, \ldots, x_n^0)$ defines the minimum of $g(x)$, subject to $f_j(x) \leqslant 0$ and to $x_i \geqslant 0$, if and only if non-negative values $y_1^0, \ldots, y_m^0$ exist which satisfy

$$\frac{\partial g}{\partial x_i}(x^0) + \sum_j y_j^0 \frac{\partial f_j}{\partial x_i}(x^0) \geqslant 0, \qquad (12\text{–}3)$$

$$\sum_i \left[\frac{\partial g}{\partial x_i}(x^0) + \sum_j y_j^0 \frac{\partial f_j}{\partial x_i}(x^0)\right] x_i^0 = 0,$$

and

$$\sum_j f_j(x^0) y_j^0 = 0. \qquad (12\text{–}4)$$

The sufficiency is easily proved. From $y_j^0 \geqslant 0$, $f_j(x) \leqslant 0$, we have $g(x) \geqslant g(x) + \sum_j y_j^0 f_j(x)$ and then from (12–1), because of the convexity of the functions g and f_j,

$$g(x) + \sum_j y_j^0 f_j(x) \geqslant g(x^0) + \sum_j y_j^0 f_j(x^0) + \sum_i \frac{\partial g}{\partial x_i}(x^0)(x_i - x_i^0)$$

$$+ \sum_j y_j^0 \sum_i \frac{\partial f_j}{\partial x_i}(x^0)(x_i - x_i^0).$$

From the two relations in (12–3) this is not smaller than

$$g(x^0) + \sum_j y_j^0 f_j(x^0),$$

and the latter equals, by (12–4), $g(x^0)$. Hence for any feasible x we have $g(x) \geqslant g(x^0)$. In other words, x^0 produces the minimum.

The proof of the necessity of the existence of a y^0 which satisfies (12–3) and (12–4) together with a x^0 that solves the minimizing problem, is somewhat more involved. In fact, it can only be given with a further assumption—always satisfied when the constraints are linear—and which will be introduced at an appropriate stage.

If the minimum is reached at a point x^0 in the interior of the feasible region, then conditions (12–3) and (12–4) are satisfied with $y_j^0 = 0$ (all j), because then

$$\frac{\partial g}{\partial x_i}(x^0) = 0 \quad \text{for all } i.$$

We turn to the case when $g(x)$ has its minimum on a boundary in point x^0, where, say, $f_1(x^0) = \ldots = f_k(x^0) = x_1^0 = \ldots = x_l^0 = 0$.

Now we introduce the additional condition (called the "constraint qualification" by Kuhn and Tucker [98]) that the feasible region is such that if for any point x^0 on the boundary defined above

$$\sum_i \frac{\partial f_j}{\partial x_i}(x^0)dx_i \leqslant 0 \quad \text{holds for } j = 1, 2, \ldots, k,$$

and

$$dx_i \geqslant 0 \quad \text{holds for } i = 1, 2, \ldots, l,$$

then the direction $dx_1, \ldots, dx_n$ is tangential to some arc from x^0 into the region.

In two dimensions the formulae mean that, from any point on the boundary, a direction which is simultaneously on the feasible sides of all tangents (i.e. forms an acute angle with their normals into the region) points into the feasible region. An example where this is not so is given in Fig. 12–8. Here the direction indicated by an arrow satisfies

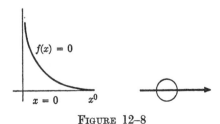

FIGURE 12–8

the condition that it is on the feasible side of both the tangents to $f(x) = 0$ and to $x = 0$ in x^0, yet we cannot move into the feasible region starting off in that direction.

If $g(x^0)$ is the minimum of $g(x)$, then we have

$$\sum_i \frac{\partial g}{\partial x_i} (x^0) dx_i \geqslant 0$$

for all directions into the interior, i.e. (by virtue of the constraints qualification), for all directions for which

$$-\sum_i \frac{\partial f_j}{\partial x_i} (x^0) dx_i \geqslant 0 \qquad (j = 1, \ldots, k),$$

and

$$dx_i \geqslant 0 \qquad (i = 1, \ldots, l).$$

It follows from Farkas' theorem (Section 3–2) that there exists a set $y_j^0 \geqslant 0$ $(j = 1, \ldots, k)$ and $z_i^0 \geqslant 0$ $(i = 1, \ldots, l)$ such that

$$\frac{\partial g}{\partial x_i} (x^0) = -\sum_{j=1}^k y_j^0 \frac{\partial f_j}{\partial x_i} (x^0) + z_i^0 \qquad (i = 1, \ldots, n)$$

where $z_{l+1}^0 = \ldots = z_n = 0$, or also a set

$$y_j^0 \geqslant 0 \quad (j = 1, \ldots, m) \qquad \text{and} \qquad z_i^0 \geqslant 0 \quad (i = 1, \ldots, n),$$

such that

$$\frac{\partial g}{\partial x_i} (x^0) = -\sum_{j=1}^m y_j^0 \frac{\partial f_j}{\partial x_i} (x^0) + z_i^0 \qquad (i = 1, \ldots, n),$$

where those y_j^0 which did not appear in the previous equation are zero.

Consequently, for all i

$$\frac{\partial g}{\partial x_i} (x^0) + \sum_{j=1}^m y_j^0 \frac{\partial f_j}{\partial x_i} (x^0) \geqslant 0,$$

and also

$$\sum_i \left[\frac{\partial g}{\partial x_i} (x^0) + \sum_{j=1}^m y_j^0 \frac{\partial f_j}{\partial x_i} (x^0) \right] x_i^0 = \sum_{i=1}^n x_i^0 z_i^0.$$

The last sum is zero, because either $x_i^0 = 0$ (for $i = 1, \ldots, l$) or $z_i^0 = 0$ (for $i = l + 1, \ldots, n$).

These two relations form (12–3). Relation (12–4) also holds, because either $f_j(x^0) = 0$ (for $j = 1, \ldots, k$) or $y_j^0 = 0$ (for $j = k + 1, \ldots, m$). This terminates the proof of the necessity of the existence of a set y^0 satisfying (12–3) and (12–4).

If the constraints are given in the form of equations, or if the variables are not restricted to non-negative values, then the conditions are modified as follows:

I. The variables need not be non-negative. Then, where Farkas' theorem comes into the argument, $dx_i \geqslant 0$ is not required; hence the $z_i^0 \geqslant 0$ do not appear, and the first relation in (12–3) becomes

$$\frac{\partial g}{\partial x_i}(x^0) + \sum_{j=1}^{m} y_j^0 \frac{\partial f_j}{\partial x_i}(x^0) = 0.$$

This implies the second relation, which need not be explicitly stated.

II. The constraints are in the form of equations, say $f_j(x) = 0$. This implies (12–4). It is now not necessary to require that the y_j be non-negative.

III. Both changes, indicated in I and in II are to be made. Thus we obtain the well-known Lagrangian condition for an extremum with side conditions. The extremum must be a minimum, because $g(x)$ is convex.

In Section 3–4 we have introduced a Lagrangian function; here we generalize the argument to the nonlinear case. Introduce, then, the *Lagrangian function*

$$\phi(x, y) = g(x) + \sum_j y_j f_j(x),$$

where the functions $g(x)$ and $f_j(x)$ are those which were defined earlier.

In terms of this function, the inequality constraints $f_j(x) \leqslant 0$, together with (12–3) and (12–4), can be written

$$\frac{\partial \phi(x, y)}{\partial x_i} \geqslant 0, \qquad \frac{\partial \phi(x, y)}{\partial y_j} \leqslant 0, \tag{12–5}$$

$$\sum_i x_i \frac{\partial \phi(x, y)}{\partial x_i} = 0, \qquad \sum_j y_j \frac{\partial \phi(x, y)}{\partial y_j} = 0. \tag{12–6}$$

Hence, if and only if non-negative values x^0 and y^0 have been found which satisfy (12–5) and (12–6) then the x^0 solve the minimizing problem.

In Section 3–4 we have seen that certain solutions x^0, together with non-negative values y^0 (which, in the linear case, solve the dual problem), satisfy the *saddle point property*

$$\phi(x^0, y) \leqslant \phi(x^0, y^0) \leqslant \phi(x, y^0).$$

Here we shall show that for this saddle point property to hold, x^0 and y^0 must satisfy (12–5) and (12–6) and that these conditions are also sufficient for a saddle point.

Our presentation is again based on [98]. Kuhn and Tucker deal with a maximizing problem, while we deal with one of minimizing. For this reason we shall adapt their argument without, of course, changing any of the essential points of it. As a consequence, we shall have proved the following.

THEOREM 12–2. If and only if non-negative values x^0, y^0 satisfy the saddle point property for

$$g(x) + \sum_j y_j f_j(x),$$

then x^0 solves the problem of minimizing $g(x)$ subject to $f_j(x) \leqslant 0$, $x_i \geqslant 0$.

Proof. (Necessity.) We want to show that values x^0 and y^0 satisfy the saddle point property only if for all i

$$\frac{\partial \phi}{\partial x_i}(x^0, y^0) \geqslant 0, \qquad \sum_i \frac{\partial \phi}{\partial x_i}(x^0, y^0)x_i^0 = 0,$$

and also for all j

$$\frac{\partial \phi}{\partial y_j}(x^0, y^0) \leqslant 0, \qquad \sum_j \frac{\partial \phi}{\partial y_j}(x^0, y^0)y_j^0 = 0.$$

Let $\phi(x, y^0)$ have its lowest value at $x_1^0, \ldots, x_n^0$. This point is either within the feasible region or on its boundary. In the former case we have (Fig. 12–9)

$$\frac{\partial \phi}{\partial x_i}(x^0, y^0) = 0 \qquad \text{for all } x_i$$

In the latter, we might have

$$\frac{\partial \phi}{\partial x_i}(x^0, y^0) > 0, \qquad \text{if } x_i^0 = 0 \text{ in that point.}$$

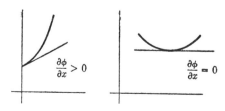

$$\frac{\partial \phi}{\partial x} > 0 \qquad\qquad \frac{\partial \phi}{\partial x} = 0$$

FIGURE 12–9

The analogous statement for y-variables follows in the same way.

(Sufficiency.) $\phi(x, y^0)$ is convex in x, and $\phi(x^0, y)$ is concave in y, so that, from (12–1) and (12–2) we have

$$\phi(x, y^0) \geqslant \phi(x^0, y^0) + \sum_i \frac{\partial \phi}{\partial x_i} (x^0, y^0)(x_i - x_i^0), \qquad (12\text{–}7)$$

and

$$\phi(x^0, y) \leqslant \phi(x^0, y^0) + \sum_j \frac{\partial \phi}{\partial y_j} (x^0, y^0)(y_j - y_j^0). \qquad (12\text{–}8)$$

Then, from (12–7) we have

$$\underline{\phi(x, y^0)} \geqslant \phi(x^0, y^0) + \sum_i \frac{\partial \phi}{\partial x_i} (x^0, y^0)(x_i - x_i^0),$$

$$\geqslant \underline{\phi(x^0, y^0)},$$

$$\geqslant \phi(x^0, y^0) + \sum_j \frac{\partial \phi}{\partial y_j} (x^0, y^0)(y_j - y_j^0),$$

from (12–8) $\geqslant \underline{\phi(x^0, y)}.$

The underlined terms are those in the saddle point condition, and the inequality signs are those required for the proof.

Following [21], we can prove, independently of more general considerations, that in the case of a convex quadratic objective function $C(x)$,

$$C(x) = c_0 + 2\sum c_i x_i + \sum\sum c_{ij} x_i x_j,$$

subject to linear equality constraints

$$\sum_i a_{ij} x_i = b_j \qquad (j = 1, \ldots, m),$$

and to

$$x_i \geqslant 0.$$

x^0 defines the minimum of $C(x)$ if y_j (not necessarily non-negative) exist satisfying conditions (12–3) which can now be written

$$2c_i + 2\sum_j c_{ij} x_j^0 + \sum_j y_j a_{ij} - v_i = 0,$$

$$\sum_i v_i x_i^0 = 0,$$

$$v_i \geqslant 0.$$

Let values $x_i \geqslant 0$ $(i = 1, \ldots, n)$ satisfying the constraints be given. Because $C(x)$ is convex we have

$$\sum_i \sum_j c_{ij} (x_i - x_i^0)(x_j - x_j^0) \geqslant 0,$$

or, by simple arithmetic,

$$\sum_i \sum_j c_{ij} x_i x_j - \sum_i \sum_j c_{ij} x_i^0 x_j^0 \geqslant 2 \sum_i \sum_j c_{ij} x_j^0 (x_i - x_i^0),$$

and therefore

$$C(x) - C(x^0) \geqslant \sum_j 2(c_i + \sum_i c_{ij} x_j^0)(x_i - x_i^0)$$

$$= \sum_i (v_i - \sum_j y_j a_{ij})(x_i - x_i^0).$$

Hence, if

$$\sum_i v_i x_i^0 = 0$$

and

$$\sum_j a_{ij} x_i = \sum_j a_{ij} x_i^0 \quad (= b_j),$$

then

$$C(x) - C(x^0) \geqslant \sum_i v_i x_i \geqslant 0.$$

This means that under the conditions stated x^0 produces the minimum of $C(x)$. We have to find non-negative x_i and v_i, together with y_j (not sign-restricted), such that

$$\sum_i a_{ij} x_i = b_j \qquad\qquad (j = 1, \ldots, m),$$

$$2(c_i + \sum_j c_{ij} x_j) + \sum_j y_j a_{ij} - v_i = 0 \qquad (i = 1, \ldots, n),$$

subject to the further conditions that

$$\sum_i v_i x_i = 0,$$

i.e. if x_1 is positive, then v_i must be zero, and vice versa. Apart from this latter condition, of a combinatorial character, this is a linear programming problem.

Note that there are now $n + m$ constraints and $2n + m$ variables. But out of the $2n$ variables x_i and v_i at least n must be zero, so that there cannot be more than $n + m$ positive variables, and this number equals that of the linear constraints. It follows that a solution of the quadratic problem is given by a basic solution of the imbedded linear programming problem. Various methods based on these ideas have been developed. For instance, Barankin and Dorfman [21] and also Frank and Wolfe [68] start off with some basic solution of the linear problem, and proceed to reduce $\sum_i x_i v_i$ to zero, while H. Markowitz [103] and P. Wolfe [127] start with weaker conditions and with $\sum_i x_i v_i = 0$ and then alter the solutions until all linear constraints are satisfied. We explain in more detail the latter procedure, due to P. Wolfe.*

* See also [83].

We assume that the constraints are not degenerate, and that all b_j are non-negative.

To begin with, we set all y_j and v_i equal to zero, and then replace the constraints by the weaker

$$\sum_i a_{ij}x_i + w_j = b_j \qquad\qquad (j = 1, \ldots, m),$$

$$2(c_i + \sum_j c_{ij}x_j) + z_{i1} - z_{i2} = 0 \qquad (i = 1, \ldots, n),$$

where the newly introduced variables w_j, z_{i1} and z_{i2} are non-negative. We minimize $\sum_j w_j$.

Having done this, we ignore the w_j and either z_{i1} or z_{i2} for each i, whichever is zero. Then we minimize

$$\sum_i{}' z_{i1} - \sum_i{}' z_{i2},$$

where the summations extend over those variables which have been retained. During this process we do not introduce x_i into the basis if v_i is in it, and vice versa.

EXAMPLE 12–3. We take the same problem as in Examples 12–1 and 12–2. First, solve the following linear programming problem:

$$2x_1 + x_2 + x_3 + w = 10,$$
$$16x_1 - 12x_2 + z_{11} - z_{12} = 44,$$
$$-12x_1 + 34x_2 + z_{21} - z_{22} = 42,$$
$$z_{31} - z_{32} = 0.$$

Minimize w.

This can be solved by inspection. The basic optimal variables are

$$x_3 = 10, \qquad z_{11} = 44, \qquad z_{21} = 42, \qquad z_{31} = 0.$$

The second phase starts with the tableau

		x_1	x_2	v_1	v_2	v_3	y	
	x_3	2	1	0	0	0	0	10
1	z_{11}	16	−12	−1	0	0	2	44
1	z_{21}	−12	34*	0	−1	0	1	42
1	z_{31}	0	0	0	0	−1	1	0
		4	22	−1	−1	−1	4	86

This problem is solved (with different notation) in Exercise 5–9. The final tableau is

	1			1	1		
	x_3	z_{21}	v_1	v_2	z_{11}	z_{31}	
x_1	0.4	−0.01	−0.005	0.01	0.005	0	3.8
v_3	−2	0.2	−0.4	−0.2	0.4	−1	6
x_2	0.2	0.02	0.01	−0.02	−0.01	0	2.4
y	−2	0.2	−0.4	−0.2	0.4	0	6
	0	−1	0	0	−1	−1	0

The values of x_i solve the problem.

Wolfe has proved that this procedure terminates after a finite number of steps, provided $C(x)$ is positive definite, or does not contain any linear terms. For the positive semidefinite case a slight modification of the method can serve. We refer the reader to the source for this, and also for the resolution of various possible complications.

We cannot do more than mention briefly one or two other methods to deal with nonlinear problems. A number of such methods are in the process of being developed, and publications will probably soon appear. An idea of J. B. Rosen (cf. [17], pp. 113–114) starts from a feasible point which satisfies at least one of the (linear) constraints as an equation. We project the gradient of the objective function into the space which is the intersection of all hyperplanes defined by such constraints; if this projection is not positive, then either the optimum has been reached (and criteria are given for this to have happened), or one of the constraints is ignored and one proceeds along the component of the gradient in the intersection space of the remaining ones. If there is a positive projection of the gradient, then one proceeds along it.

To decide how far to proceed, Rosen suggests an interpolation between the point considered and the farthest feasible point in the direction decided on. This interpolation produces the point at which the component of the gradient would be zero if it changed linearly along the edge, as it does when the objective function is quadratic. G. Zoutendijk [128] has developed three methods which differ by the direction in which one proceeds from a feasible point. Given such a point, draw around it an infinitesimal sphere, or an infinitesimal cuboid, and proceed in the direction of the point which is highest in this region. In a third method, also using the cuboid, those sides of it which have lower points than the centre are ignored.

An algorithm for a nonlinear transportation problem was given by E. M. L. Beale in [27]. More advanced problems are dealt with in [2].

12–5 Quadratic duality. There exist theorems in nonlinear programming which deal with properties analogous to Duality relationships in the linear case, and we shall now exhibit results due to W. S. Dorn, concerning quadratic programming. (See [58].)

Consider the following problem:

$$\text{Minimize } \tfrac{1}{2}\sum_i\sum_j c_{ij}x_i x_j + \sum_i c_i x_i = f(x), \text{ say,}$$

where the double sum is a positive semidefinite form and $c_{ij} = c_{ji}$, subject to

$$\sum_i a_{ij}x_i \geqslant b_j \qquad (j = 1, \ldots, m),$$

$$x_i \geqslant 0 \qquad (i = 1, \ldots, n),$$

and let $(x_1^0, \ldots, x_n^0)$ be an optimal solution. Then this is also an optimal solution of the linear programming problem:

$$\text{Minimize } \sum_i\sum_j c_{ij}x_i x_j^0 + \sum_i c_i x_i = F(x),$$

subject to the same constraints. For suppose there exists a solution $(x_1', \ldots, x_n')$ of the constraints such that

$$F(x') < F(x^0);$$

then

$$x_i'' = x_i^0 + k(x_i' - x_i^0) \qquad (i = 1, \ldots, n)$$

is also a solution of the constraints, and k could be chosen so as to make

$$f(x'') < f(x^0).$$

But this would contradict the optimality of $(x_1^0, \ldots, x_n^0)$ for the quadratic problem.

The dual to the linear problem just considered is:

$$\text{Maximize } \sum_j b_j y_j = G(y), \text{ say,}$$

subject to

$$\sum_j a_{ij}y_j \leqslant \sum_j c_{ij}x_j^0 + c_i \qquad (i = 1, \ldots, n),$$

$$y_j \geqslant 0 \qquad (j = 1, \ldots, m).$$

Denote the optimizing solution of the latter by $(y_1^0, \ldots, y_m^0)$. Of course $G(y^0) = F(x^0)$.

We introduce now a quadratic problem, which we call the dual to the previous quadratic problem, as follows:

$$\text{Maximize } -\tfrac{1}{2}\sum_i\sum_j c_{ij}u_i u_j + \sum_j b_j y_j = g(u, y), \text{ say,}$$

subject to

$$\sum_j a_{ij} y_j - \sum_j c_{ij} u_j \leqslant c_i,$$

$$y_j \geqslant 0 \qquad (j = 1, \ldots, m).$$

THEOREM. The optima of the two objective functions $f(x)$ and $g(u, y)$ are equal.

Proof. It is immediately seen that $u_i = x_i^0$, $y_j = y_j^0$ is a feasible solution of the constraints. Let u_i', y_j' be another feasible solution. We can then show that

$$g(x^0, y^0) - g(u', y') \geqslant 0.$$

We have

$$g(x^0, y^0) - g(u', y')$$
$$= \tfrac{1}{2}\left(\sum\sum c_{ij} u_i' u_j' - \sum\sum c_{ij} x_i^0 x_j^0\right) + \sum_j b_j y_j^0 - \sum_j b_j y_j'.$$

Now

$$\sum_i \sum_j c_{ij}(u_i' - x_i^0)(u_j' - x_j^0) \geqslant 0,$$

because the c_{ij} are, by assumption, coefficients of a positive semi-definite form. Hence

$$\sum_i \sum_j c_{ij} u_i' u_j' - \sum_i \sum_j c_{ij} x_i^0 x_j^0 \geqslant 2\sum_i \sum_j c_{ij} x_i^0 (u_j' - x_j^0).$$

Consequently

$$g(x^0, y^0) - g(u', y') \geqslant \sum_i \sum_j c_{ij} x_i^0 (u_j' - x_j^0) + \sum_j b_j y_j^0 - \sum_j b_j y_j'.$$

By the Duality theorem of Linear Programming, the right-hand side equals

$$\sum_i \sum_j c_{ij} x_i^0 (u_j' - x_j^0) + \sum_i \sum_j c_{ij} x_i^0 x_j^0 + \sum_i c_i x_i^0 - \sum_j b_j y_j',$$

while, from the constraints of the dual problems, we have

$$\sum_i \sum_j c_{ij} x_i^0 u_j' \geqslant \sum_i x_i^0 \left(\sum_j a_{ij} y_j' - c_i\right),$$

and

$$-\sum_j b_j y_j' \geqslant -\sum_i \sum_j a_{ij} x_i^0 y_j',$$

so that, after substitution and reduction, we have

$$g(x^0, y^0) - g(u', y') \geqslant 0.$$

Also

$$g(x^0, y^0) = -\tfrac{1}{2}\sum_i\sum_j c_{ij}x_i^0x_j^0 + \sum_j b_j y_j^0$$
$$= -\tfrac{1}{2}\sum_i\sum_j c_{ij}x_i^0x_j^0 + \sum_i\sum_j c_{ij}x_i^0x_j^0 + \sum_i c_i x_i^0$$
$$= f(x^0). \qquad \text{Q.E.D.}$$

Dorn has also shown that, conversely, if u_i^0 and y_j^0 form an optimal solution of the second quadratic problem, then $x_i = u_i^0$ is an optimal solution of the first.

J. R. Dennis, in [56], has proved a similar theorem concerning positive definite quadratic forms, in which case either problem can be considered as the primal, and the other as its dual.

EXERCISES

12–1. Minimize

$$4t_0^2 + 4t_0t_1 + 6t_1^2 + 4t_0 - 56t_1 + 180,$$

subject to

$$t_0 \geqslant -3, \qquad t_1 \leqslant 6.$$

12–2. Minimize

$$x_1^2 + x_2^2 - 4x_1 - 2x_2 + 5,$$

subject to

$$x_1 + x_2 \leqslant 4, \qquad x_1, x_2 \geqslant 0.$$

12–3. Which of the following quadratic forms are positive definite, positive semidefinite, or neither? Give examples to illustrate your answers.

(a) $5x_1^2 + 2x_2^2 + 2x_3^2 - 2x_1x_2 - 2x_1x_3 + 4x_2x_3$

(b) $5x_1^2 + 3x_2^2 + 2x_3^2 - 2x_1x_2 - 2x_1x_3 + 4x_2x_3$

(c) $5x_1^2 + x_2^2 + 2x_3^2 - 2x_1x_2 - 2x_1x_3 + 4x_2x_3.$

12–4. Construct and solve the problem which is dual to Example 12–1.

CHAPTER 13

DYNAMIC PROGRAMMING

This chapter deals with an approach to programming which is useful where time, or a sequence of decisions, is essential. The theory is almost entirely due to Richard Bellman and his school, and we can here only deal with a few examples which show the spirit of the method. For more detail we refer the reader to [3].

13–1 Principle of optimality. The procedures of Dynamic Programming are inspired by the *Principle of Optimality*: An optimal policy has the property that, whatever the initial state and the initial decision, the remaining decisions must constitute an optimal policy with regard to the state resulting from the first decision.

For an application of this principle, consider the linear programming problem

$$\sum_{i=1}^{n} a_{ij}x_i \leqslant b_j \qquad (j = 1, \ldots, m),$$

$$x_i \geqslant 0 \qquad (i = 1, \ldots, n).$$

$$\text{Minimize } \sum_i c_i x_i = C, \text{ say.}$$

The a_{ij} are all supposed to be non-negative.

Denote the minimum of C subject to the constraints above, and to

$$x_{k+1} = \ldots = x_n = 0,$$

by $C_k(b_1, \ldots, b_m)$. If x_n is given, say $x_n = \bar{x}_n$, then

$$C = c_1 x_1 + \ldots + c_{n-1}x_{n-1} + c_n \bar{x}_n = \bar{C}, \text{ say,}$$

subject to

$$\sum_{i=1}^{n-1} a_{ij}x_i \leqslant b_j - a_{nj}\bar{x}_n \qquad (j = 1, \ldots, m).$$

The modified linear programming problem, which requires to minimize $\bar{C}$ subject to the latter constraints, has the minimal objective function

$$C_{n-1}(b_1 - a_{n1}\bar{x}_n, \ldots, b_m - a_{nm}\bar{x}_n) + c_n \bar{x}_n,$$

and therefore the minimum we are looking for is found by choosing an appropriate $\bar{x}_n$, remembering that $b_j - a_{nj}\bar{x}_n \geqslant 0$. Hence

$$C_n(b_1, \ldots, b_m) = \min [c_n x_n + C_{n-1}(b_1 - a_{n1}x_n, \ldots, b_m - a_{nm}x_n)],$$

where x_n is restricted to lie in the range

$$0 \leqslant x_n \leqslant \min_j b_j/a_{nj}.$$

For computational purposes there is not much to be gained from such a general equation, but in some cases its solution can be found by an iterative process.

As a first illustration, take the following case. We have n machines at our disposal and they can do two jobs. If z of them do the first job, they produce goods worth $g(z)$, and if z do the second job, they produce goods worth $h(z)$. The machines are subject to attrition, so that after doing the first job only $a(z)$ out of z remain available, and if they were doing the second job, then this available number is $b(z)$. Here a, b, g, and h are given functions. The process is repeated with the remaining machines, and it is required to maximize the total return of N stages.

Let x_1 machines do the first job, and $y_1 = n - x_1$ the second. The return is $g(x_1) + h(n - x_1)$, and the remaining available machines are $a(x_1) + b(n - x_1)$. This amount is divided into x_2 and y_2, and thus we continue. The total return after N stages is

$$\sum_{i=1}^{N} [g(x_i) + h(y_i)],$$

subject to

$$x_i + y_i = n_i \qquad\qquad (n_1 = n),$$
$$n_{i+1} = a(x_i) + b(y_i) \qquad (i = 1, \ldots, N - 1),$$
$$0 \leqslant x_i \leqslant n_i \qquad\qquad (i = 1, \ldots, N).$$

If the functions are linear, then we have here a linear programming problem; but the dynamic programming approach is helpful in other cases as well.

Let $f_N(n)$ be the maximum total return obtainable from an N-stage process starting with an amount n of machines. Then

$$f_1(n) = \max_{0 \leqslant x \leqslant n} [g(x) + h(n - x)],$$

and

$$f_k(n) = \max_{0 \leqslant x \leqslant n} \{g(x) + h(n - x) + f_{k-1}[a(x) + b(n - x)]\}$$

for $k > 1$.

Starting from $f_1(n')$, where n' is the number of machines available at the beginning of the Nth stage, we work backwards to $f_N(n)$.

EXAMPLE 13-1. If

$$N = 3, \quad a(x) = 0.3x, \quad b(y) = 0.6y, \quad g(x) = 0.8x, \quad h(y) = 0.5y,$$

then

$$f_1(n) = \max_{0 \leqslant x \leqslant n} (0.5n + 0.3x).$$

Whatever the value of n, the maximum of the linear function $0.5n + 0.3x$ is obtained for $x = n$. Hence $f_1(n) = 0.8n$, and

$$f_1[a(x) + b(n - x)] = f_1[0.3x + 0.6(n - x)] = 0.48n - 0.24x.$$

Continuing, we have

$$f_2(n) = \max_{0 \leqslant x \leqslant n} [0.8x + 0.5(n - x) + 0.48n - 0.24x]$$

$$= \max_{0 \leqslant x \leqslant n} (0.98n + 0.06x) = 1.04n,$$

which is obtained from $x = n$. Then

$$f_3(n) = \max_{0 \leqslant x \leqslant n} \{0.8x + 0.5(n - x) + 1.04[0.3x + 0.6(n - x)]\}$$

$$= \max_{0 \leqslant x \leqslant n} (1.124n - 0.012x) = 1.124n,$$

which is obtained from $x = 0$. The optimal schedule is therefore that shown in Table 13-1.

TABLE 13-1

Stage	First job	Second job	Result	Remaining machines
1	0	n	$0.5n$	$0.6n$
2	$0.6n$	0	$0.48n$	$0.18n$
3	$0.18n$	0	$0.144n$	
			$1.124n$	

When we formulate this example in linear programming terms, we have:

Maximize $0.3(x_1 + x_2 + x_3) + 0.5(n_1 + n_2 + n_3) = B$,

subject to $0 \leqslant x_i \leqslant n_i$ $(i = 1, 2, 3)$;

$$n_2 = 0.6n_1 - 0.3x_1, \qquad n_3 = 0.6n_2 - 0.3x_2.$$

This is easily reduced to:

Maximize $0.98n + 0.06x_1 + 0.15x_2 + 0.3x_3 = B$,

subject to

$$x_1 \leqslant n, \qquad x_2 - n_2 \leqslant 0, \qquad x_3 - n_3 \leqslant 0,$$
$$n_2 + 0.3x_1 = 0.6n,$$
$$n_3 + 0.18x_1 + 0.3x_2 = 0.36n.$$

This problem is solved in Exercise 5–7 and in Exercise 10–5 for $n = 15$, in integers.

We shall now show that if g and h are both convex functions, then an optimal policy requires that at each stage the variable take either its upper or its lower limit, provided $g(0) = h(0) = 0$. Indeed, when the two functions are convex, then $g(x) + h(n - x)$ is also convex as a function of x, and hence $f_1(n)$ equals either $g(n)$ or $h(n)$, whichever is larger, because a convex function takes its maximum value at a limit of its range.

Hence $f_1(n)$ is also a convex function (of n), and so is $f_2(n)$, the latter being the larger of $g(n) + f_1[a(n)]$ and $h(n) + f_1[b(n)]$; so we continue to $f_N(n)$.

An even simpler problem, that of "Distribution of Effort", was considered by A. Vazsonyi in [119]. Let the effect of an effort x expended on a job i be given by the function $f_i(x)$, and let the upper limit of all available efforts be n. It is required to maximize the total effect

$$\sum_{i=1}^{N} f_i(x_i),$$

subject to

$$\sum_{i=1}^{N} x_i \leqslant n, \qquad x_i \geqslant 0 \quad (i = 1, \ldots, N).$$

Denote the maximum of

$$\sum_{i=1}^{m} f_i(x_i),$$

subject to

$$\sum_{i=1}^{m} x_i \leqslant a, \qquad x_i \geqslant 0 \quad (i = 1, \ldots, m),$$

by $F_m(a)$; then we have

$$F_m(n) = \max_{0 \leqslant x \leqslant n} [F_{m-1}(n - x) + f_m(x)],$$

and the problem can again be solved by a method of recurrence.

13-2 Functional equations. The essential feature of the dynamic programming approach is the reduction of a multivariable problem to a succession of single variable cases. This results in deciding at each stage on the values of these single variables, and their sequence defines a *policy*. The optimum policy looked for is that which yields the maximum of an objective function, e.g. that which yields $f_N(n)$ in Example 13-1.

If we consider the connection between $f_k(n)$ and $f_{k-1}(n)$, we may be justified to argue, in appropriate situations, that for large k the two functions are the same. Then we obtain a functional equation which would, for instance, in Example 13-1, read

$$f(n) = \max_{0 \leqslant x \leqslant n} \{g(x) + h(n - x) + f[a(x) + b(n - x)]\}.$$

The solution of such a functional equation, if available, can then give a clue to approximations to the solution of the original problem. It is for this reason that dynamic programming theory is also concerned with existence and uniqueness theorems about solutions of functional equations. In some cases the solution to such an equation provides, in fact, the precise answer (see the example below, quoted from [30]).

To give an idea of the scope of this theory, we mention a few more examples of its applications.

The topic of Exercise 6–7 has been dealt with in a paper by Richard Bellman, in [32]. A paper by the same author [30] deals with the following problem, where no subscript such as k above enters: Let

$f(t)$ be the over-all discounted return from a machine of age t when an optimal policy is applied;

$r(t)$ the discounted value of the future output of a machine of that age;

$u(t)$ the discounted future cost of its upkeep; and

$s(t)$ its price on the market when sold.

We can then at each age of the machine decide whether we want to sell it and replace it by a new one, or whether we want to keep it. A new machine costs p.

In the first case we obtain

$$s(t) - p + r(0) - u(0) + af(1),$$

where a is the discounting factor for one year; in the second case we obtain

$$r(t) - u(t) + af(t + 1).$$

Hence $f(t)$ equals the larger of these two expressions. The solution of the resulting functional equation is discussed in the paper mentioned above.

Finally, to show the value of Dynamic Programming in a problem of pure mathematics, we mention the following from [31]: Prove that the maximum of $x_1 x_2 \ldots x_n$, subject to

$$x_i \geqslant 0, \qquad \sum_i x_i = 1,$$

is $(1/n)^n$. It follows then that the maximum of the same product, subject to $x_i \geqslant 0$, is

$$\left[\sum_i \frac{x_i}{n} \right]^n,$$

whence

$$(x_1 x_2 \ldots x_n)^{1/n} \leqslant \sum_i \frac{x_i}{n},$$

a well-known inequality connecting the geometric and arithmetic means.

Proof. Let
$$f_k(a) = \max x_{n-k+1} \ldots x_n,$$
subject to
$$x_{n-k+1} + \ldots + x_n = a > 0, \qquad x_i \geqslant 0.$$

We have then the recurrence relations

$$f_1(a) = a,$$
$$f_{k+1}(a) = \max_{0 \leqslant x_{n-k} \leqslant a} [x_{n-k} f_k(a - x_{n-k})].$$

This is solved by
$$f_k(a) = a^k (1/k)^k,$$

because this is the maximum, for $0 \leqslant x \leqslant a$, of

$$x(a - x)^{k-1} \left(\frac{1}{k-1} \right)^{k-1}.$$

EXERCISES

13-1. Solve Example 13–1 for the following numerical data:

$$N = 3, \qquad a = \tfrac{1}{3}, \qquad b = \tfrac{2}{3}, \qquad g(x) = 3x, \qquad h(y) = 2.5y.$$

13-2. A merchant can buy a number of items at 10 each, and resells them at 25 each. He can return those which he has not sold after one month and recover 5 for each, and he can also buy new items, again at 10 each. At the end of two months he can return unsold items for 2 each. Determine his best policy: How many should he buy to begin with, or after one month? How many should he return after one month?

Assume the following probability distribution of demands during the first two months:

First month: required number of items

	0	1	2	3	4	5
Probability	1/8	1/8	1/4	1/4	1/8	1/8.

Second month: required number of items

	0	1	2
Probability	1/4	1/2	1/4.

It is required to optimize the expected profit.

13-3. Solve Example 4-1 by an application of the Principle of Optimality.

13-4. A man who can swim at a speed c wants to cross a river. At a distance x from the opposite bank the speed of the river's flow is v_x. It is required to determine in which direction to swim at any given stage so as to make the deviation downstream from the point opposite to the start as small as possible.

Formulate the functional equation for solving this problem, assuming that v_x is constant for all values above $x - 1$ up to and including x. Solve the functional equation for $v_x = x$ (a constant larger than c).

APPENDIX

Consider the set of linear equations

$$a_{1j}x_1 + \ldots + a_{Nj}x_N = b_j \qquad (j = 1, \ldots, m).$$

The pattern of coefficients

$$\begin{pmatrix} a_{11} & \cdots & a_{N1} \\ \cdot & & \\ \cdot & & \\ \cdot & & \\ a_{1m} & \cdots & a_{Nm} \end{pmatrix}$$

is called a *m-by-N matrix.* The *N*-by-*m* matrix

$$\begin{pmatrix} a_{11} & \cdots & a_{1m} \\ \cdot & & \\ \cdot & & \\ \cdot & & \\ a_{N1} & \cdots & a_{Nm} \end{pmatrix}$$

is called its *transpose.*

A 1-by-N matrix is a *row-vector,* and its transpose, a N-by-1 matrix, a *column-vector.*

Two matrices can be multiplied, if the number of columns of the first factor equals that of the rows of the second. The *product* of the two matrices

$$\begin{pmatrix} a_{11} & \cdots & a_{N1} \\ \cdot & & \\ \cdot & & \\ \cdot & & \\ a_{1m} & \cdots & a_{Nm} \end{pmatrix} \quad \text{and} \quad \begin{pmatrix} b_{11} & \cdots & b_{n1} \\ \cdot & & \\ \cdot & & \\ \cdot & & \\ b_{1N} & \cdots & b_{nN} \end{pmatrix}$$

is defined as the matrix (n-by-m)

$$\begin{pmatrix} a_{11}b_{11} + a_{21}b_{12} + \ldots + a_{N1}b_{1N} & \cdots & a_{11}b_{n1} + a_{21}b_{n2} + \ldots + a_{N1}b_{nN} \\ \cdot & & \\ \cdot & & \\ \cdot & & \\ a_{1m}b_{11} + a_{2m}b_{12} + \ldots + a_{Nm}b_{1N} & \cdots & a_{1m}b_{n1} + a_{2m}b_{n2} + \ldots + a_{Nm}b_{nN} \end{pmatrix}.$$

As a special case, take $n = m = 1$. The result is the *inner product* of two vectors, and for convenience the second is often also written as a row: $(a_1, \ldots, a_N) \cdot (b_1, \ldots, b_N) = a_1b_1 + \ldots + a_Nb_N$. Thus this inner product reduces to a single number. Two vectors whose inner product is zero are called *orthogonal*, by analogy with the geometric meaning in two or three dimensions.

If both matrices are square (i.e. the number of columns equals the number of rows), then the two matrices can be multiplied, whichever is the first factor, but the two products are not necessarily the same.

The square matrix

$$\begin{pmatrix} 1 & 0 & \dots & 0 \\ 0 & 1 & \dots & 0 \\ \cdot & & & \\ \cdot & & & \\ \cdot & & & \\ 0 & 0 & \dots & 1 \end{pmatrix}$$

is called the *unit matrix*. If we multiply a given matrix M by it, the result is again M, whether the unit matrix is the first or the second factor, i.e. whether we *premultiply* or *postmultiply* by it.

If the product of two matrices is the unit matrix, then we call them *inverse* to one another. On exchanging their order as factors, the result remains the unit matrix.

A column with a single 1, and all other entries 0 is a *unit vector*.

To a square n-by-n matrix we attach a number, called its *determinant*, which we write

$$\begin{vmatrix} a_{11} & \dots & a_{n1} \\ \cdot & & \\ \cdot & & \\ \cdot & & \\ a_{1n} & \dots & a_{nn} \end{vmatrix}.$$

The *order* of this determinant is n. Its value is defined as follows: When $n = 2$, then

$$\begin{vmatrix} a_{11} & a_{21} \\ a_{12} & a_{22} \end{vmatrix} = a_{11}a_{22} - a_{21}a_{12}.$$

For $n > 2$, we have the recurrence relation

$$\begin{vmatrix} a_{11} & \dots & a_{n1} \\ \cdot & & \\ \cdot & & \\ \cdot & & \\ a_{1n} & \dots & a_{nn} \end{vmatrix} = a_{11}\begin{vmatrix} a_{22} & \dots & a_{n2} \\ \cdot & & \\ \cdot & & \\ \cdot & & \\ a_{2n} & \dots & a_{nn} \end{vmatrix} - a_{21}\begin{vmatrix} a_{12} & \dots & u_{n2} \\ \cdot & & \\ \cdot & & \\ a_{1n} & \dots & a_{nn} \end{vmatrix}$$

$$+ \dots + (-1)^{n-1}a_{n1}\begin{vmatrix} a_{12} & \dots & a_{n-1.2} \\ \cdot & & \\ \cdot & & \\ a_{1n} & \dots & a_{n-1.n} \end{vmatrix}.$$

The determinants on the right-hand side with their signs affixed are called *co-factors*, of $a_{11}, \ldots, a_{n1}$ respectively.

The transpose of a matrix has the same determinant. An exchange of the positions of two rows, or of two columns, results in a change of sign of the determinant. Hence the recurrence relation above—a special case of the so-called *Laplace expansion*—can be applied to other rows than the first, and to columns as well. It follows, also, that if two rows or two columns are equal, or if a whole row or a whole column consists entirely of zeros, then the determinant is zero.

A matrix consisting of a portion of the original matrix is called a *submatrix* of the latter, and the determinant of a square submatrix is a *minor*.

The *rank* of a matrix is defined as the order of the minor of highest order which is not zero. A square n-by-n matrix whose rank is smaller than n is called *singular*: its determinant is zero.

A system of linear equations

$$a_{1j}x_1 + \ldots + a_{Nj}x_N = b_j \qquad (j = 1, \ldots, m)$$

has a solution if the rank of the matrix of coefficients equals that of the *extended matrix of coefficients*:

$$\begin{pmatrix} a_{11} & \cdots & a_{N1} & b_1 \\ \cdot & & & \\ \cdot & & & \\ \cdot & & & \\ a_{1m} & \cdots & a_{Nm} & b_m \end{pmatrix}.$$

If, in particular, the rank of both matrices is $N = m = n$, say, then the system of linear equations is solved by (*Cramer's rule*)

$$x_1 = \begin{vmatrix} b_1 & a_{21} & \cdots & a_{n1} \\ \cdot & & & \\ \cdot & & & \\ \cdot & & & \\ b_n & a_{2n} & \cdots & a_{nn} \end{vmatrix} \div \begin{vmatrix} a_{11} & a_{21} & \cdots & a_{n1} \\ \cdot & & & \\ \cdot & & & \\ a_{1n} & a_{2n} & \cdots & a_{nn} \end{vmatrix}$$

$$\cdot$$
$$\cdot$$
$$\cdot$$

$$x_n = \begin{vmatrix} a_{11} & a_{21} & \cdots & b_1 \\ \cdot & & & \\ \cdot & & & \\ \cdot & & & \\ a_{1n} & a_{2n} & \cdots & b_n \end{vmatrix} \div \begin{vmatrix} a_{11} & a_{21} & \cdots & a_{n1} \\ \cdot & & & \\ \cdot & & & \\ a_{1n} & a_{2n} & \cdots & a_{nn} \end{vmatrix}.$$

If all $b_j = 0$, but the rank of the matrix of coefficients is n, then the only solution of the set is the *trivial solution* $x_1 = x_2 = \ldots = x_n = 0$.

If the matrix is singular, then there exists an infinity of solutions. Clearly, given some solution, we can multiply each variable by a constant, and the result is again a solution.

If the rank of a matrix

$$\begin{pmatrix} a_{11} & \cdots & a_{N1} \\ \cdot & & \\ \cdot & & \\ \cdot & & \\ a_{1m} & \cdots & a_{Nm} \end{pmatrix}$$

is r, say, and $r < N$, then it contains a nonsingular r-by-r matrix, say

$$\begin{pmatrix} a_{11} & \cdots & a_{r1} \\ \cdot & & \\ \cdot & & \\ \cdot & & \\ a_{1r} & \cdots & a_{rr} \end{pmatrix},$$

and we can express the jth $(j = r + 1, \ldots, N)$ column of the original matrix as a *linear combination* of the first r columns, i.e. as follows:

$$a_{j1} = t_1 a_{11} + \ldots + t_r a_{r1},$$
$$\cdot$$
$$\cdot$$
$$\cdot$$
$$a_{jm} = t_1 a_{1m} + \ldots + t_r a_{rm}.$$

Similarly, if $r < m$, we can express the ith $(i = r + 1, \ldots, m)$ row as a linear combination of the first r rows. Hence, if in a system of m equations in N variables both the rank of the matrix of coefficients, and that of the extended matrix of coefficients, is $r < m$, then $m - r$ equations are linear combinations of r of them. We say then that the equations are *linearly dependent*. If $r = m$, they are *linearly independent*.

SOLUTIONS TO EXERCISES

CHAPTER 2

2–1. This is a Transportation Problem, as can be seen by adding the equation $x_3 + x_6 = 3$. It has therefore $3^1 \cdot 2^2 = 12$ basic solutions. The variables x_2, x_3, x_5, x_6 do not form a set of basic variables, because the matrix of their coefficients is singular. The same is true of x_1, x_3, x_4, x_6 and of x_1, x_2, x_4, x_5. All other sets of four variables are basic. The determinants of their coefficients are either 1 or -1. (This is always the case in a transportation problem; see Section 4–8.)

The simplest way of exhibiting all basic solutions is to construct tables where we enter the values of the variables, such that entries in rows add up to the right-hand sides of the first two equations, and those in columns to the right-hand sides of the last three. Thus

	3	3	3
4	x_{11}	x_{12}	x_{13}
5	x_{21}	x_{22}	x_{23}

The basic solutions are then the following (only the values of the basic variables are entered):

-2	3	3
5		

3	-2	3
5		

3	3	-2
		5

1	3	
2		3

3	1	
	2	3

1		3
2	3	

3		1
	3	2

4		
-1	3	3

	1	3
3	2	

	3	1
3	2	

	4	
3	-1	3

		4
3	3	-1

The feasible solutions are, of course, those where no entry is negative.

2–2. The following are the basic solutions, none of them feasible:

0   $x_3 = 1$, $x_4 = -6$; a $x_1 = 1$, $x_4 = -3$;

b $x_1 = 2$, $x_3 = -1$; c $x_2 = 1$, $x_4 = -4$;

d $x_2 = 3$, $x_3 = -2$; e $x_1 = 4$, $x_2 = -3$.

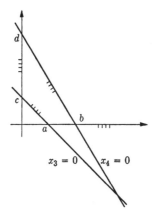

2–3. To the basic solutions listed for Exercise 2–2, the following are now added: (A) $x_1 = 1$, $x_5 = 3$; (B) $x_2 = 1$, $x_5 = 4$; (C) $x_3 = 1$, $x_5 = 6$. The

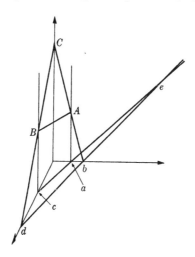

variables x_4, x_5 do not form a basic set, since the axis $x_1 = x_2 = 0$ and the plane $x_3 = 0$ have no point in common.

The feasible region is that above the triangle ABC. No point below it is feasible. This explains why Exercise 2–2 had no feasible solution at all.

2–4. Basic solutions:

(a) $x_1 = 3$, $x_2 = 0$; $x_1 = 0$, $x_2 = 6$.

(b)

Basic Variables		Nonbasic
$x_1 = 3$,	$z_1 = 0$	x_2, z_2
$x_1 = 3$,	$z_2 = 0$	x_2, z_1
$x_2 = 6$,	$z_1 = 0$	x_1, z_2
$x_2 = 6$,	$z_2 = 0$	x_1, z_1.

Setting $z_1 = z_2 = 0$ does not produce a basic solution.

2–5. The matrix

$$\begin{pmatrix} 1 & 2 & 0 \\ 1 & 2 & 0 \\ 1 & 1 & 1 \end{pmatrix}$$

is singular. Hence setting x_3, x_4, and x_5 equal to zero does not produce a vertex. We can solve the last two equations for x_1 and x_2; thus

$$x_1 = 2 - \frac{5x_3}{3} + x_5 - 2x_6,$$

$$x_2 = \frac{2x_3}{3} - x_5 + x_6.$$

The variable x_6, which is now equal to $\frac{1}{2}$, can be increased to 1 or decreased to 0. The vertices obtained thereby are

$$(0, 1, 0, 0, 0, 1) \quad \text{and} \quad (2, 0, 0, 0, 0, 0)$$
$$\text{(basic variables: } x_2, x_3, x_6 \quad x_1, x_2, x_3),$$

and the given point lies half-way between them. (Compare Example 2–7 and Fig. 2–4.)

CHAPTER 3

3–1. Because only $y_1 = y_2 = 0$ satisfy the second group of constraints, we must have $x_i > 0$, which solve the first group. For instance, $x_1 = 2$, $x_2 = x_3 = 1$.

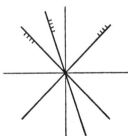

3–2. The only non-negative values x_i which satisfy $3x_1 + x_2 = 0$ are $x_1 = x_2 = 0$, so that we must find y_1 and y_2 such that

$$3y_1 - 2y_2 > 0, \qquad y_1 + 2y_2 > 0.$$

For instance, $y_1 = y_2 = 1$. [*Note*: A construction on the lines of the proof of Theorem 3–1 would produce $y_1 = 288$, $y_2 = 16$.]

3–3. We have three inequalities and three variables. We solve therefore the system

$$3t_1 + 5t_2 + 4t_3 = 22,$$
$$2t_1 - 3t_2 - t_3 = 1,$$
$$-t_1 + t_2 + 5t_3 = 8.$$

We obtain $t_1 = 3$, $t_2 = 1$, $t_3 = 2$. These values are all positive, and thus the statement of the exercise is proved. The coefficients looked for are the values of t_i just given.

3–4. The system

$$3t_1 + 5t_2 + 4t_3 = 12$$
$$2t_1 - 3t_2 - t_3 = 7$$
$$-t_1 + t_2 + 5t_3 = 6$$

is solved (only) by $t_1 = 3$, $t_2 = -1$ (negative!), $t_3 = 2$. An example satisfying the three inequalities, but not $12x_1 + 7x_2 + 6x_3 \geqslant 0$ is $x_1 = 1$, $x_2 = -1$, $x_3 = -1$.

CHAPTER 4

4–1. The tabular labelling is

$\oplus$	a	b	c	d	e	f	$\ominus$	
$\oplus$		36	15	47	0	0	0	0	
a	0		19	0	0	0	0	40	$(36, \oplus)$
b	0	19		0	28	0	0	0	$(15, \oplus)$
c	0	0	0		20	34	0	0	$(47, \oplus)$
d	0	0	28	20		. 0	21	0	
e	0	0	0	34	0		14	38	
f	0	0	0	0	21	14		24	
$\ominus$	0	0	0	0	0	0	0		$(36, a)$ [*contd.*]

⊕	a	b	c	d	e	f	⊖	
⊕		0	15	47	0	0	0	0	
a	36		19	0	0	0	0	4	(15, b)
b	0	19		0	28	0	0	0	(15, ⊕)
c	0	0	0		20	34	0	0	(47, ⊕)
d	0	0	28	20		0	21	0	(15, b)
e	0	0	0	34	0		14	38	(34, c)
f	0	0	0	0	21	14		24	(15, d)
⊖	0	36	0	0	0	0	0		(15, f)

⊕	a	b	c	d	e	f	⊖	
⊕		0	0	47	0	0	0	0	
a	36		19	0	0	0	0	4	
b	15	19		0	13	0	0	0	(20, b)
c	0	0	0		20	34	0	0	(47, ⊕)
d	0	0	43	20		0	6	0	(20, c)
e	0	0	0	34	0		14	38	(34, c)
f	0	0	0	0	36	14		9	(6, d)
⊖	0	36	0	0	0	0	15		(34, e)

⊕	a	b	c	d	e	f	⊖	
⊕		0	0	13	0	0	0	0	
a	36		19	0	0	0	0	4	(13, a)
b	15	19		0	13	0	0	0	(13, d)
c	34	0	0		20	0	0	0	(13, ⊕)
d	0	0	43	20		0	6	0	(13, c)
e	0	0	0	68	0		14	4	(6, f)
f	0	0	0	0	36	14		9	(6, d)
⊕	0	36	0	0	0	34	15		(6, f)

⊕	a	b	c	d	e	f	⊖	
⊕		0	0	7	0	0	0	0	
a	36		19	0	0	0	0	4	(7, b)
b	15	19		0	13	0	0	0	(7, d)
c	40	0	0		14	0	0	0	(7, ⊕)
d	0	0	43	26		0	0	0	(7, c)
e	0	0	0	68	0		14	0	
f	0	0	0	0	42	14		3	
⊖	0	36	0	0	0	34	21		(4, a)

	⊕	a	b	c	d	e	f	⊖	
⊕		0	0	3	0	0	0	0	
a	36		23	0	0	0	0	0	(3, b)
b	15	15		0	17	0	0	0	(3, d)
c	44	0	0		10	0	0	0	(3, ⊕)
d	0	0	39	30		0	0	0	(3, c)
e	0	0	0	68	0		14	4	
f	0	0	0	0	42	14		3	
⊖	0	40	0	0	0	34	21		

4–2. There is no such set of distinct representatives. The first six sets contain between them only the five numbers 1–5. According to König's theorem it must therefore be possible to cover all crosses in the following matrix by less than eight lines:

	1	2	3	4	5	6	7	8	9
(a)	×	×	×	×	×				
(b)		×	×	×					
(c)	×		×	×	×				
(d)	×		×	×					
(e)			×	×					
(f)		×		×					
(g)	×			×		×			×
(h)		×			×		×	×	×

On the other hand, if we added a number 7 to the set (f), then eight lines would be necessary, and a DSR would be 1,2, 5, 4, 3, 7, 6, 8.

4–3. The shortest path from A to B has length 7 (by inspection). The dual graph is shown in part (a) of the figure shown on the following page. A maximum flow, viz. 7, is indicated. It can be determined as shown in part (b).

4–4. Consider the table as a determinant and develop it into its terms (ignoring the + or − signs). The subscripts in the nonzero terms indicate the positions of the independent "1"s. They are

21	12	53	34	45		21	32	13	44	55		21	32	53	14	45
31	12	23	44	55		21	42	13	34	55		31	42	23	14	55
51	12	23	34	45		51	32	13	44	25		51	32	23	14	45
31	12	53	44	25		51	42	13	34	25		31	42	53	14	25.

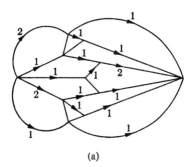

(a)

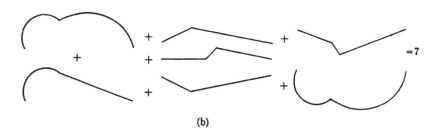

(b)

CHAPTER 5

5–1.

(a)

	1 x_1	-1 x_2	
x_3	1	1	3
x_4	1	-2	1
x_5	-2	1*	2
	-1	1	0

	1 x_1	x_5	
x_3	3*	-1	1
x_4	-3	2	5
x_2	-2	1	2
	1	-1	-2

(with -1 left of x_2)

	x_3	x_5	
x_1	1/3	$-1/3$	1/3
x_4	1	1	6
x_2	2/3	1/3	8/3
	$-1/3$	$-2/3$	$-7/3$

(with 1 left of x_1, -1 left of x_2)

(b)

	1 x_1	-1 x_2	
x_3	1	1	3
x_4	1*	-2	1
x_5	-2	1	2
	-1	1	0

	-1 x_4	x_2	
x_3	-1	3*	2
x_1	1	-2	1
x_5	2	-3	4
	1	-1	1

(with 1 left of x_1)

	x_4	x_3	
x_2	$-1/3$	1/3	2/3
x_1	1/3	2/3	7/3
x_5	1	1	6
	2/3	1/3	5/3

(with -1 left of x_2, 1 left of x_1)

5-2. Omit $x_1 + x_2 \leqslant 3$. Then no pivot is available in either of the second tableaux.

5-3. $t > 2$ produces the minimum at $x_1 = x_5 = 0$. [See second tableau for problem (a).] $t < \frac{1}{2}$ produces the maximum at $x_2 = x_4 = 0$. [See second tableau for problem (b).] $t = 2$, and $t = \frac{1}{2}$ give straight-line contours parallel to $x_5 = 0$ and to $x_4 = 0$, respectively.

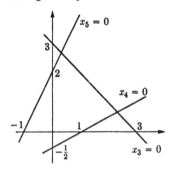

5-4.

(a)

		2	−1	−1	
		x_1	x_2	x_3	
−M	x_{101}	−1	2	−1	1
−M	x_{102}	−1	−1	2*	1
		− 2	1	1	0
	(M)	2	−1	−1	−2

		2	−1	
		x_1	x_2	
−M	x_{101}	−3/2	3/2*	3/2
−1	x_3	−1/2	−1/2	1/2
		−3/2	3/2	−1/2
	(M)	3/2	−3/2	−3/2

		2	
		x_1	
−1	x_2	−1	1
−1	x_3	−1	1
		0	−2

This is also the final tableau for problem (b). The value -2 is the minimum as well as the maximum of the objective function. In fact,
$$2x_1 - x_2 - x_3 = -(-x_1 + 2x_2 - x_3) - (-x_1 - x_2 + 2x_3) = -2 \text{ (constant)}.$$

The final tableau discloses the fact that the two constraints are equivalent to

$$x_2 = 1 + x_1, \qquad x_3 = 1 + x_1,$$

where x_1 has an arbitrary non-negative value.

5-5. Denote the slack variables by $z_1, \ldots, z_6$. We have then the following three successive tableaux:

		11	10	
		x	y	
	z_1	-0.5	1.3^*	0.8
	z_2	4	1	10.7
	z_3	6	1	15.4
	z_4	6	-1	13.4
	z_5	4	-1	8.7
	z_6	5	-3	10.0
	B	-11	-10	0.0

		11		
		x	z_1	
10	y	-0.3846	0.7692	0.6154
	z_2	4.3846^*	-0.7692	10.0846
	z_3	6.3846	-0.7692	14.7846
	z_4	5.6154	0.7692	14.0154
	z_5	3.6154	0.7692	9.3154
	z_6	3.8462	2.3077	11.8462
	B	-14.8462	7.6923	6.1538

		z_2	z_1	
10	y	0.0877	0.7017	1.5
11	x	0.2281	-0.1754	2.3
	z_3	-1.4561	0.3508	0.1
	z_4	-1.2807	1.7543	1.1
	z_5	-0.8246	1.4035	1.0
	z_6	-0.8772	2.9824	3.0
	B	3.3860	5.0877	40.3

Note that we have chosen as the pivotal column in the first tableau that with the *least* negative shadow price. If we had followed the more usual procedure, then it would have required six exchanges to obtain the final tableau. The basic variables would then have been as follows:

$B =$ 0.0, $z_1 = 0.8,$ $z_2 = 10.7,$ $z_3 = 15.4,$ $z_4 = 13.4,$ $z_5 = 8.7,$ $z_6 = 10.0$

     22.0   1.8   2.7   3.4   1.4   0.7 $x =$ 2.0

     30.3   1.3   1.0   1.1   0.1 $y =$ 0.5   2.3

     32.85   1.065   0.6   0.6 $z_6 =$ 0.35   0.7   2.35

     36.4   0.7   0.1 $z_5 =$ 0.1   1.0   1.0   2.4

     38.85   0.285 $z_4 =$ 0.6   0.6   2.15   1.3   2.35

     40.3 $z_3 =$ 0.1   1.1   1.0   3.0   1.5   2.3

Of course, we could not have known this. The graph explains how it has come about.

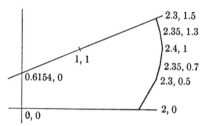

5–6. Eliminate x_8, substituting for it $183 - 44x_1 - 42x_2 + x_4$. Hence

$$2x_1 + x_2 + x_3 \qquad\qquad = 10,$$
$$80x_1 - 60x_2 \qquad - x_4 + x_5 \qquad = 200,$$
$$-120x_1 + 340x_2 \qquad - x_4 \qquad + x_6 \qquad = 1700,$$
$$32x_1 + 36x_2 \qquad - x_4 \qquad\qquad + x_7 = 104.$$

Minimize $183 - 44x_1 - 42x_2 + x_4$.

	-44	-42	1	
	x_1	x_2	x_4	
x_3	2	1	0	10
x_5	80*	-60	-1	200
x_6	-120	340	-1	1700
x_7	32	36	-1	104
x_8	44	42	-1	183

		-42	1	
	x_5	x_2	x_4	
x_3	$-1/40$	$5/2$	$1/40$	5
-44 x_1	$1/80$	$-3/4$	$-1/80$	$5/2$
x_6	$3/2$	250	$-5/2$	2000
x_7	$-2/5$	60*	$-3/5$	24
x_8	$-11/20$	75	$-9/20$	73

		x_5	x_7	1 x_4	
	x_3	$-1/120$	$-1/24$	$1/20^*$	4
-44	x_1	$3/400$	$1/80$	$-1/50$	$14/5$
	x_6	$19/6$	$-25/6$	0	1900
-42	x_2	$-1/150$	$1/60$	$-1/100$	$2/5$
	x_8	$-1/20$	$-5/4$	$3/10$	43

		x_5	x_7	x_3	
1	x_4	$-1/6$	$-5/6$	20	80
-44	x_1	$1/240$	$-1/240$	$2/5$	$22/5$
	x_6	$19/6^*$	$-25/6$	0	1900
-42	x_2	$-1/120$	$1/120$	$1/5$	$6/5$
	x_8	0	-1	-6	19

An optimal tableau

		x_6	x_7	x_3	
1	x_4	$1/19$	$-20/19$	20	180
-44	x_1	$-1/760$	$1/760$	$2/5$	$19/10$
	x_5	$6/19^*$	$-25/19$	0	600
-42	x_2	$1/380$	$-1/380$	$1/5$	$31/5$
	x_8	0	-1	-6	19

Another optimal tableau

5–7. Let y_i be slack variables, and z_j artificial variables.

		x_1	x_2	x_3	n_2	n_3	
	y_1	1^*	0	0	0	0	n
	y_2	0	1	0	-1	0	0
	y_3	0	0	1	0	-1	0
1	z_1	0.3	0	0	1	0	$0.6n$
1	z_2	0.18	0.3	0	0	1	$0.36n$
	B	-0.06	-0.15	-0.3	0	0	$0.98n$
	C'	0.48	0.3	0	1	1	$0.96n$

		y_1	x_2	x_3	n_2	n_3	
	x_1	1	0	0	0	0	n
	y_2	0	1*	0	-1	0	0
	y_3	0	0	1	0	-1	0
1	z_1	-0.3	0	0	1	0	$0.3n$
1	z_2	-0.18	0.3	0	0	1	$0.18n$
	B	0.06	-0.15	-0.3	0	0	$1.04n$
	C'	-0.48	0.3	0	1	1	$0.48n$

		y_1	y_2	x_3	n_2	n_3	
	x_1	1	0	0	0	0	n
	x_2	0	1	0	-1	0	0
	y_3	0	0	1	0	-1	0
1	z_1	-0.3	0	0	1*	0	$0.3n$
1	z_2	-0.18	-0.3	0	0.3	1	$0.18n$
	B	0.06	0.15	-0.3	-0.15	0	$1.04n$
	C'	-0.48	-0.3	0	1.3	1	$0.48n$

					1		
		y_1	y_2	x_3	z_1	n_3	
	x_1	1	0	0	0	0	n
	x_2	-0.3	1	0	1	0	$0.3n$
	y_3	0	0	1	0	-1	0
	n_2	-0.3	0	0	1	0	$0.3n$
1	z_2	-0.09	-0.3	0	-0.3	1*	$0.09n$
	B	0.015	0.15	-0.3	0.15	0	$1.085n$
	C'	-0.09	-0.3	0	-1.3	1	$0.09n$

					1	1	
		y_1	y_2	x_3	z_1	z_2	
	x_1	1	0	0	0	0	n
	x_2	-0.3	1	0	1	0	$0.3n$
	y_3	-0.09	-0.3	1	-0.3	1	$0.09n$
	n_2	-0.3	0	0	1	0	$0.3n$
	n_3	-0.09	-0.3	0	-0.3	1	$0.09n$
	B	0.015	0.15	-0.3	0.15	0	$1.085n$
	C'	0	0	0	-1	-1	0

End of phase I

		y_1	y_2	0.3 x_3	
0.06	x_1	1	0	0	n
0.15	x_2	-0.3	1	0	$0.3n$
	y_3	-0.09	-0.3	1^*	$0.09n$
	n_2	-0.3	0	0	$0.3n$
	n_3	-0.09	-0.3	0	$0.09n$
B		0.015	0.15	-0.3	$1.085n$

		y_1	y_2	y_3	
0.06	x_1	1^*	0	0	n
0.15	x_2	-0.3	1	0	$0.3n$
0.3	x_3	-0.09	-0.3	1	$0.09n$
	n_2	-0.3	0	0	$0.3n$
	n_3	-0.09	-0.3	0	$0.09n$
B		-0.012	0.06	0.3	$1.112n$

		0.06 x_1	y_2	y_3	
	y_1	1	0	0	n
0.15	x_2	0.3	1	0	$0.6n$
0.3	x_3	0.09	-0.3	1	$0.18n$
	n_2	0.3	0	0	$0.6n$
	n_3	0.09	-0.3	0	$0.18n$
B		0.012	0.06	0.3	$1.124n$

5–8. Twice the first, plus the second equation, equals the third. Hence there are only two independent equations. They can equivalently be written

$$x_1 - x_2 = 1, \qquad x_3 - x_4 = 1.$$

Take x_1 and x_3 as the first basic variables. This gives

		1 x_2	1 x_4	
1	x_1	-1	0	1
1	x_3	0	-1	1
		-2	-2	2

,

and this is, also the final tableau.

If we had not noticed that the equations were dependent, then we could have proceeded as follows:

		1	1	1	1	
		x_1	x_2	x_3	x_4	
M	x_{101}	1	−1	1	−1	2
M	x_{102}	2*	−2	−1	1	1
M	x_{103}	4	−4	1	−1	5
		−1	−1	−1	−1	0
	(M)	7	−7	1	−1	8

		1	1	1	
		x_2	x_3	x_4	
M	x_{101}	0	3/2*	−3/2	3/2
1	x_1	−1	−1/2	1/2	1/2
M	x_{103}	0	3	−3	3
		−2	−3/2	−1/2	1/2
	(M)	0	9/2	−9/2	9/2

		1	1	
		x_2	x_4	
1	x_3	0	−1	1
1	x_1	−1	0	1
M	x_{103}	0	0	0
		−2	−2	2

which is the same final tableau as before.

5–9. The constraints are as follows:

$$2x_1 + x_2 \leqslant 10,$$
$$16x_1 - 12x_2 - x_3 + 2x_6 \leqslant 44,$$
$$-12x_1 + 34x_2 - x_4 + x_6 \leqslant 42,$$
$$-x_5 + x_6 \leqslant 0,$$

and the objective function is

$$86 - 4x_1 - 22x_2 + x_3 + x_4 + x_5 - 4x_6.$$

The minimizing problem leads to the following tableaux:

	x_1	x_9	x_3	x_4	x_5	x_6	
x_7	40/17	−1/34	0	1/34	0	−1/34	149/17
x_8	200/17	6/17	−1	−6/17	0	40/17	1000/17
x_2	−6/17	1/34	0	−1/34	0	1/34	21/17
x_{10}	0	0	0	0	−1	1*	0
C	200/17	−11/17	−1	−6/17	−1	57/17	1000/17

	x_1	x_9	x_3	x_4	x_5	x_{10}	
x_7	40/17*	−1/34	0	1/34	−1/34	1/34	149/17
x_8	200/17	6/17	−1	−6/17	40/17	−40/17	1000/17
x_2	−6/17	1/34	0	−1/34	1/34	−1/34	21/17
x_6	0	0	0	0	−1	1	0
C	200/17	−11/17	−1	−6/17	40/17	−57/17	1000/17

	x_7	x_9	x_3	x_4	x_5	x_{10}	
x_1	17/40	−1/80	0	1/80	−1/80	1/80	149/40
x_8	−5	1/2	−1	−1/2	5/2*	−5/2	15
x_2	3/20	1/40	0	−1/40	1/40	−1/40	51/20
x_6	0	0	0	0	−1	1	0
C	−5	−1/2	−1	−1/2	5/2	−7/2	15

	x_7	x_9	x_3	x_4	x_8	x_{10}	
x_1	0.4	−0.01	−0.005	0.01	0.005	0	3.8
x_5	−2	0.2	−0.4	−0.2	0.4	−1	6
x_2	0.2	0.02	0.01	−0.02	−0.01	0	2.4
x_6	−2	0.2	−0.4	−0.2	0.4	0	6
C	0	−1	0	0	−1	−1	0

5–10.

		x_1	x_2	
1	z_1	2*	1	6
1	z_2	4	2	12
	C	6	3	18

		z_1	x_2	
1	x_1	1/2	1/2*	3
1	z_2	−2	0	0
	C	−3	0	0

		z_1	x_1	
1	x_2	1	2	6
1	z_2	−2	0	0
	C	−3	0	0

Two basic solutions

Compare this with Exercise 2–4.

5-11. (a) Minimize $3y_3 + y_4 + 2y_5 + My_{100}$. The equations

$$y_3 + y_4 - 2y_5 - y_1 = 1,$$
$$y_3 - 2y_4 + y_5 - y_2 = -1$$

can be combined into

$$3y_4 - 3y_5 - y_1 + y_2 = 2.$$

This replaces the second equation and leads to the following tableaux:

			3	1	2	
		y_1	y_3	y_4	y_5	
	y_2	-1	0	3	-3	2
M	y_{100}	-1	1^*	1	-2	1
	C	0	-3	-1	-2	0
	(M)	-1	1	1	-2	1

			1	2	
		y_1	y_4	y_5	
	y_2	-1	3^*	-3	2
3	y_3	-1	1	-2	1
	C	-3	2	-8	3

				2	
		y_1	y_2	y_5	
1	y_4	$-1/3$	$1/3$	-1	$2/3$
3	y_3	$-2/3$	$-1/3$	-1	$1/3$
	C	$-7/3$	$-2/3$	-6	$5/3$

For problem (b)

		3	1	2	
	y_3	y_4	y_5		
y_1	-1	-1^*	2	-1	
y_2	-1	2	-1	1	
C	-3	-1	-2	0	

			3		2	
		y_3	y_1	y_5		
1	y_4	1	-1	-2	1	
	y_2	-3^*	2	3	-1	
	C	-2	-1	-4	1	

				2	
		y_2	y_1	y_5	
1	y_4	$1/3$	$-1/3$	-1	$2/3$
3	y_3	$-1/3$	$-2/3$	-1	$1/3$
	C	$-2/3$	$-7/3$	-6	$5/3$

5-12. (a) The succession of inverse matrices $D_{u_1 \ldots u_m}$ is as follows (the slack variables are denoted by $y_i (i = 1, \ldots, 5)$:

							$z_{u,0}$	z_{u,x_1}
y_1	1	0	0	0	0	0	1	12
y_2	0	1	0	0	0	0	1	7
y_3	0	0	1	0	0	0	1	14
y_4	0	0	0	1	0	0	1	5
y_5	0	0	0	0	1	0	1	16*
	0	0	0	0	0	1	0	−1

Shadow prices:	x_1	x_2	x_3	x_4	x_5
	−1	−1	−1	−1	−1

							$z_{u,0}$	z_{u,x_4}
y_1	1	0	0	0	−0.75	0	0.25	4.25
y_2	0	1	0	0	−0.4375	0	0.5625	15.5625*
y_3	0	0	1	0	−0.875	0	0.125	2.125
y_4	0	0	0	1	−0.3125	0	0.6875	17.6875
x_1	0	0	0	0	0.0625	0	0.0625	0.0625
	0	0	0	0	0.0625	1	0.0625	−0.9375

Shadow prices:	y_5	x_2	x_3	x_4	x_5
	0.0625	−0.8125	0.125	−0.9375	0.25

							$z_{u,0}$	z_{u,x_5}
y_1	1	−0.2731	0	0	−0.6305	0	0.0964	2.5703
x_4	0	0.0643	0	0	−0.0281	0	0.0361	−0.3695
y_3	0	−0.1365	1	0	−0.8153	0	0.0482	1.2851
y_4	0	−1.1365	0	1	0.1847	0	0.0482	1.2851*
x_1	0	−0.0040	0	0	0.0643	0	0.0602	1.2731
	0	0.0602	0	0	0.0361	1	0.0964	−0.0964

Shadow prices:	y_2	y_5	x_2	x_3	x_5
	0.0602	0.0361	−0.0482	−0.0482	−0.0964

							$z_{u,0}$	z_{u,y_2}
y_1	1	2	0	−2	−1	0	0	2*
x_4	0	−0.2625	0	0.2875	0.025	0	0.05	−0.2625
y_3	0	1	1	−1	−1	0	0	1
x_5	0	−0.8844	0	0.7781	0.14375	0	0.0375	−0.8845
x_1	0	1.1219	0	−0.9906	−0.11875	0	0.0125	1.1219
	0	−0.025	0	0.075	0.05	1	0.1	−0.025

Shadow prices:	y_2	y_4	y_5	x_2	x_3
	−0.025	0.075	0.05	0	0

							z_{u_s0}	$z_{u_sx_2}$
y_2	0.5	1	0	-1	-0.5	0	0	0
x_4	0.13125	0	0	0.025	-0.10625	0	0.05	1*
y_3	-0.5	0	1	0	-0.5	0	0	0
x_5	0.4422	0	0	-0.10625	-0.2984	0	0.0375	0.5
x_1	-0.5609	0	0	0.13125	0.4422	0	0.0125	-0.5
	0.0125	0	0	0.05	0.0375	1	0.1	0

Shadow prices:

	y_1	y_4	y_5	x_2	x_3
	0.0125	0.05	0.0375	0	0

This is one of the optimal tableaux. There are others, and they will now be determined. The shadow prices (for their definition, see section 7–1) remain unaltered. The last row and the last column of $D_{u_1 \ldots u_m}$ will be omitted in what follows.

						z_{u_s0}	$z_{u_sx_3}$
y_2	0.5	1	0	-1	-0.5	0	0
x_2	0.13125	0	0	0.025	-0.10625	0.05	0
y_3	-0.5	0	1	0	-0.5	0	0
x_5	0.3766	0	0	-0.11575	-0.2453	0.0125	0.5*
x_1	-0.4953	0	0	0.14375	0.3891	0.0375	0.5

						z_{u_s0}	$z_{u_sx_4}$
y_2	0.5	1	0	-1	-0.5	0	0
x_2	0.13125	0	0	0.025	-0.10625	0.05	1
y_3	-0.5	0	1	0	-0.5	0	0
x_3	0.7531	0	0	-0.2375	-0.4906	0.025	-1
x_1	-0.8719	0	0	0.2625	0.6344	0.025	1*

						z_{u_s0}	$z_{u_sx_5}$
y_2	0.5	1	0	-1	-0.5	0	1
x_2	1.0031	0	0	-0.2375	-0.7406	0.025	1*
y_3	-0.5	0	1	0	-0.5	0	0
x_3	-0.11875	0	0	0.025	0.14375	0.05	1
x_4	-0.8719	0	0	0.2625	0.6344	0.025	-1

						z_{u_s0}	$z_{u_sx_1}$
y_2	0.5	1	0	-1	-0.5	0	0
x_5	1.0031	0	0	-0.2375	-0.7406	0.025	-1
y_3	-0.5	0	1	0	-0.5	0	0
x_3	-1.1219	0	0	0.2625	0.8844	0.025	2*
x_4	0.13125	0	0	0.025	-0.10625	0.05	0

This leads back to the first optimal solution. We have now found all optimal solutions to the problem.

(b) Inspection shows that all dual solutions are given by exchanging x_i and y_i, for all i.

(c) The solutions to the problem with four main variables are those above, ignoring solutions which contain x_5. The solutions to the problem with three main variables are again those above, ignoring solutions which contain either x_4 or x_5; only one solution remains, viz.

$$x_1 = 0.025, \qquad x_2 = 0.05, \qquad x_3 = 0.025.$$

The only solution to the problem

$$12x_1 + 7x_2 \leqslant 1,$$
$$7x_1 + 14x_2 \leqslant 1.$$

Maximize: $x_1 + x_2 = B$, say,

is $x_1 = 7/119$, $x_2 = 5/119$, $B = 12/119$. That of its dual is $y_1 = 7/119$, $y_2 = 5/119$.

CHAPTER 6

6–1.

	1	2	3	4	5	6	Shadow cost	
1	0	㉟*	0	31	0	0	35	
2	㉟*	0	㉟	22	30	0	35	(1)
3	㉛	22	0	0	㉛*	26	31	(5)
4	⑭	0	0	10	0	0	14	
5	0	㉚	16	29	0	㉚*	30	
6	0	0	16	26	㊱	0	36	(0)
Shadow cost	0	0	0	0	0	0		
	(3)		(2)		(6)			

	1	2	3	4	5	6	Shadow cost	
1	0	㉟*	0	31	0	0	35	
2	㉟	0	㉟*	22	30	0	35	
3	㉛*	22	0	0	㉛	26	31	(1)
4	⑭	0	0	10	0	0	14	(0)
5	0	㉚	16	29	0	㉚*	30	
6	0	0	16	26	㊱*	0	36	(5)
Shadow cost	0	0	0	0	0	0		
	(4)				(3)			

$t = 4.$

1	0	㉟*	0	31	0	0	35
2	35	0	㉟*	22	30	0	35
3	㉛*	22	0	0	㉛	26	27
4	⑭	0	0	⑩*	0	0	10
5	0	㉚	16	29	0	㉚*	30
6	0	0	16	26	㊱*	0	32
Shadow cost	4	0	0	0	4	0	

6–2.

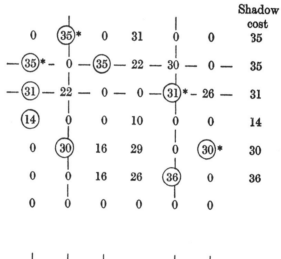

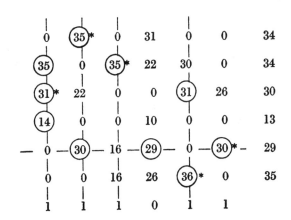

$t = 1.$

						Shadow cost
0	㉟*	0	31	0	0	34
㉟	0	㉟*	22	30	0	34
㉛*	22	0	0	㉛	26	30
⑭	0	0	10	0	0	13
— 0 —㉚— 16 —㉙— 0 —㉚*—						29
0	0	16	26	㊱*	0	35
1	1	1	0	1	1	

$t = 3.$

0	㉟*	0	㉛	0	0	31
㉟	0	㉟*	22	30	0	31
㉛*	22	0	0	㉛	26	27
⑭	0	0	⑩*	0	0	10
0	㉚	16	㉙	0	㉚*	29
0	0	16	26	㊱*	0	32
4	4	4	0	4	1	

6-3. The solution is

	$1 + x$	$1 + x$	$1 + x$	$1 + x$
$1 + 4x$	①$+ x$			$3x$
1			①	
1		①		
1		x	x	①$- 2x$

[*Note*: We want to minimize, hence $u_i + v_i \geqslant r_{ij}$ is required.]

6-4. The exercise is solved as follows

Cost table

	B	D	H	L	M	S	Y		B	D	H	L	M	S	Y
B	0	15	M	19	36	M	M	B	9.99	0.01					
D	15	0	47	28	M	20	M	D		10					
H	M	47	0	M	M	34	38	H			10				
L	19	28	M	0	40	21	24	L				8.97		1.02	0.01
M	36	M	M	40	0	M	M	M	0.02			1.04	10.01		
S	M	20	34	21	M	0	14	S				1.01		8.99	
Y	M	M	38	24	M	14	0	Y							10

M indicates not existent paths.

After rounding, and elimination of the diagonal entries, we obtain the route M-L-S-H.

6-5. The transportation problem is solved by the following final table:

	2	3	5	3	2	Shadow cost
1	1					0
7	1	3	0	3		−1
7			5		2	−1
Shadow cost	8	6	6	4	6	

or, also, by

	2	3	5	3	2	Shadow cost
1	1					0
7	1	3		3		−1
7			0	5	2	2
Shadow cost	8	6	3	4	3.	

The shadow costs are those constants which are to be subtracted from the corresponding rows and columns. Adding some constant to all row costs and subtracting it from all column costs would produce an alternative answer.

6-6. Let $x_{ikt} = 1$ if job i is done in the kth period on the tth machine, and $= 0$ otherwise. $(i = 1, \ldots, 6; \; k = 1, 2, 3; \; t = 1, 2.)$ All jobs must be done some time on some machine; hence

$$\sum_k \sum_t x_{ikt} = 1 \qquad \text{(all } i\text{)}.$$

The tth machine must be busy in all periods; hence

$$\sum_i x_{ikt} = 1 \qquad \text{(all } k, t\text{)}.$$

Also, $x_{ikt} = 0$ or 1.

The latter condition is redundant, if we can transform the others into those of a transportation problem. This is done by introducing a subscript j for the pairs (k, t) as follows:

$$(k, t) = (1, 1)\ (1, 2)\ (2, 1)\ (2, 2)\ (3, 1)\ (3, 2).$$
$$j\ \ =\ \ 1\qquad 2\qquad 3\qquad 4\qquad 5\qquad 6$$

We have then the transportation problem for $x_{ij}(i, j = 1, \ldots, 6)$, with the cost table

	$j =$						
	1	2	3	4	5	6	
1	①*	2	①	3	2	3	(1)
2	2	3	3	①*	①	2	
3	3	①*	2	2	3	①	(2)
4	2	3	3	2	①*	3	(5)
5	①	2	3	2	①	3	(0)
6	3	①	①*	3	2	2	(3)
	(5)	(6)	(1)		(5)	(3)	

We solve this by the Hungarian method, or by labelling, remembering that we want to *maximize*. We have already indicated the first stage, for either method. It leads at once to

	1	2	3	4	5	6
1	○		○ *			
2				○ *	○	
3		○				○ *
4					○ *	
5	○ *					○
6		○ *	○			

i.e. to

	9–11	1–3	5–7
1st machine	job 5	job 1	job 4
2nd machine	job 6	job 2	job 3.

In the present case an even simpler procedure is available which shows also that the solution obtained is unique. If only costs 1 are used, then the fourth job must be given to $j = 5$, after that the second job to $j = 4$, the fifth to $j = 1$, and so on until all are allocated.

6–7. We consider the store of new napkins, and the napkins used on one day and sent to the laundry, or into a store of soiled napkins, as sources, and the napkins required on a particular day, or those in the final store of soiled napkins, as destinations. We can then construct a transportation problem, provided we know an upper bound to the number of new napkins to be bought. It is easily seen that the largest of the sums of three consecutive requirements will not be exceeded by the number of napkins to be bought. This largest sum is 270. Hence we have the problem:

Cost table   **Requirement table**

									120	60	70	100	90	70	110	270

30	30	30	30	30	30	30	0	270
	15	15	5	5	5	5	0	120
		15	15	5	5	5	0	60
			15	15	5	5	0	70
				15	15	5	0	100
					15	15	0	90
						15	0	70
							0	110

The cost in the lower left-hand triangle is very large in all cells. A solution is

	120	60	70	100	90	70	110	270	Shadow cost
270	120	60	70					20	0
120				100	20				−20
60					60				−20
70						50	20		−10
100					10		90		−10
90						20		70	0
70								70	0
110								110	0.
Shadow cost	30	30	30	25	25	15	15	0	

This means:

1st day: Buy 120. After use, send them to the slow service.

2nd day: Buy 60. Send them to the slow service.

3rd day: Buy 70. Send them to the slow service.

4th day: Use 100 of those that were sent to the slow service on the first day.

5th day: Use 20, 60, and 10 respectively of those sent to the slow service on the first and second day, and to the fast service on the fourth day.

6th day: Use 50 of those sent to the slow service on the third day, and 20 of those sent to the fast service on the fifth day.

7th day: Use 20 and 90 respectively of those sent to the slow service on the third and on the fourth day.

6–8. A basic solution is the following, with the shadow costs indicated on the margins.

```
.   .   .   1   0   .       0
1   .   .   .   .   0      38
.   .   .   .   .   1      23
.   .   1   .   0   .      20
.   1   .   .   0   .       6
0   .   .   .   1   .      −2

5   5   7   11  19  10
```

Precisely the same shadow costs apply to the following alternative solution:

```
.   .   .   1   0   .
.   .   .   .   .   1
.   1   .   .   .   0
.   .   1   .   0   .
.   0   .   .   1   .
1   .   .   .   0   .
```

[The table is taken from T. E. Easterfield, "A Combinatorial Algorithm," *Journal of the London Mathematical Society*, **21**, 219–226 (1946).]

6–9. One optimal solution (with shadow costs) is

	15	20	30	35	Shadow cost
25				25	0
25	15	0		10	−1
50		20	30		0
Shadow cost	9	3	4	7	Total cost 535.

Another optimal solution is

	15	20	30	35	Shadow cost
25				25	0
25		15		10	−1
50	15	5	30		0
Shadow cost	9	3	4	7	Total cost 535.

Both these (basic) solutions were derived by Hitchcock by a straight-forward application of the algebraic procedure implicit in the Simplex method, without the use of dual variables (shadow costs).

6–10. The following tables shows the availabilities and requirements in the marginal totals, and in the upper left-hand corner of each cell its capacity limitation. The other entry in a cell is its unit cost.

	3	5	5
2	1/ 6	1/ 1	2/ 1
2	2/ 5	1/ 2	2/ 7
9	3/ 9	5/ 7	4/ 8

Some of the limitations are redundant, because they are implied by a marginal total. They could be replaced by any higher number without changing the problem.

The present problem can be transformed into an ordinary one as follows:

	3	5	5	2	3	3
1	6	M	M	0	M	M
1	M	1	M	0	M	M
2	M	M	1	0	M	M
2	5	M	M	M	0	M
1	M	2	M	M	0	M
2	M	M	7	M	0	M
3	9	M	M	M	M	0
5	M	7	M	M	M	0
4	M	M	8	M	M	0

Because many partial sums of column totals equal partial sums of row totals, we use a device against degeneracy, and obtain the following optimal solution:

	$3 + x$	$5 + x$	$5 + x$	$2 + x$	$3 + x$	$3 + x$	Shadow cost
$1 + 6x$				$1 + 6x$			0
1		$5x$		$1 - 5x$			0
2			2				-1
2	$1 - x$				$1 + x$		2
1		1					1
2					2		2
3	$2 + 2x$					$1 - 2x$	6
5		$4 - 4x$				$1 + 4x$	6
4				$3 + x$		$1 - x$	6
Shadow cost	3	1	2	0	-2	-6	

Ignoring x, and compressing the table into a solution of the original capacitated problem, we obtain

	3	5	5
2			2
2	1	1	
9	2	4	3

CHAPTER 7

7–1. (a) The dual to the problem of Exercise 5–1 is solved as Exercise 5–11. When the coefficient of y_4 in the objective function is changed from 1 into 2, the final tableau remains optimal, and hence the value of the objective function increases by 2/3 (the shadow cost in the final tableau of Exercise 5–1), and the value of y_4 in the final tableau of Exercise 5–11.

(b) When the coefficient is changed into 4, then the tableau ceases to be final, and another iteration is required:

		2			
		y_2	y_1	y_5	
4	y_4	$1/3^*$	$-1/3$	-1	$2/3$
3	y_5	$-1/3$	$-2/3$	-1	$1/3$
	C	$1/3$	$-10/3$	-9	$11/3$

		4	2		
		y_4	y_1	y_5	
	y_2	3	-1	-3	2
3	y_5	1	-1	-2	1
	C	-1	-3	-8	3

The final value of the objective function is 3. This is an increase, from 5/3, of 4/3. Thus an increase of 1 in the coefficient results in an increase of 2/3 in the objective function, but an increase of 3 does not result in an increase of three times that amount, but only in twice that amount.

7-2. The final tableau of the original problem is (see Example 7-1)

		x_4	x_5	2 x_3	
3	x_1	0.6	-0.8	-1	4
6	x_2	-0.2	0.6	1	2
	B	0.6	1.2	1	24

This agrees with

$$20 \times \quad 0.6 + 10 \times (-0.8) = 4,$$
$$20 \times (-0.2) + 10 \times \quad 0.6 = 2,$$
$$20 \times \quad 0.6 + 10 \times \quad 1.2 = 24.$$

If now c_1 ($= 20$) or c_2 ($= 10$) are changed, into 30 or 20 respectively, then new tableaux can be constructed in which only the columns of the values of the variables and of B are different; thus

(a)

		x_4	x_5	2 x_3	
3	x_1	0.6	-0.8	-1	10
6	x_2	-0.2	0.6	1	0
	B	0.6	1.2	1	30
		(30)	(10)		

This is still an optimal tableau, and thus we are prepared to pay up to $30 - 24 = 6$, i.e. 10 times 0.6 for 10 additional yards of red wool.

(b)

		x_4	x_5	2 x_3	
3	x_1	0.6	-0.8	-1^*	-4
6	x_2	-0.2	0.6	.1	8
	B	0.6	1.2	1	36
		(20)	(20)		

This is not feasible; hence we apply the Dual Simplex method (see next page):

				3	
		x_4	x_5	x_1	
2	x_3	−0.6	0.8	−1	4
6	x_2	0.4	−0.2	1	4
	B	1.2	0.4	1	32

We are prepared to pay, for 10 additional yards of green wool, up to $32 - 24 = 8$, which is less than 10 times 1.2 (shadow cost of x_5).

7-3. It is required to find values x_i and y_j such that

$$3x_1 + 2x_2 + 3x_3 \leqslant 12,$$
$$4x_1 + \ x_2 + 2x_3 \leqslant 15,$$
$$3y_4 + 4y_5 \geqslant 5,$$
$$2y_4 + \ y_5 \geqslant 3,$$
$$3y_4 + 2y_5 \geqslant 4,$$

and $12y_4 + 15y_5 = 5x_1 + 3x_2 + 4x_3$. This is equivalent to solving one of two dual problems, either maximizing $5x_1 + 3x_2 + 4x_3$ or minimizing $12y_1 + 15y_2$:

		5	3	4	
		x_1	x_2	x_3	
	x_4	3	2	3	12
	x_5	4*	1	2	15
		−5	−3	−4	0

			3	4	
		x_5	x_2	x_3	
	x_4	−3/4	5/4*	3/2	3/4
5	x_1	1/4	1/4	1/2	15/4
		5/4	−7/4	−3/2	$18\frac{3}{4}$

				4	
		x_5	x_4	x_3	
3	x_2	−3/5	4/5	6/5	3/5
5	x_1	2/5	−1/5	1/5	18/5
		1/5	7/5	3/5	$19\frac{4}{5}$

Hence the imputed costs are 7/5 for a and 1/5 for b.

CHAPTER 9

9–1. The procedure is as follows:

		$1+t$	$-1+t$	
		x_1	x_2	
	x_3	1	1	3
	x_4	1*	-2	1
	x_5	-2	1	2
	B	$-1-t$	$1-t$	0

This is a maximum for $t \leqslant -1$.

			$-1+t$	
		x_4	x_2	
	x_3	-1	3*	2
$1+t$	x_1	1	-2	1
	x_5	2	-3	4
	B	$1+t$	$-1-3t$	$1+t$

This is a maximum for $-1 \leqslant t \leqslant -\frac{1}{3}$.

		x_4	x_3	
$-1+t$	x_2	$-1/3$	$1/3$	$2/3$
$1+t$	x_1	$1/3$	$2/3$	$7/3$
	x_5	1	1	6
	B	$2/3$	$1/3+t$	$5/3+3t$

This is a maximum for $t \geqslant -1/3$.

The subscript of the basic variables are

3, 4, 5	1, 3, 5	1, 2, 5

t ——————|———————|—————
-1 $-\frac{1}{3}$

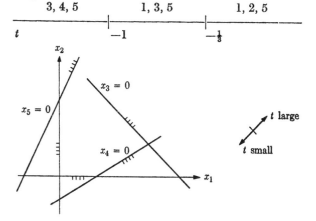

Compare this with Exercise 5–3, where the parameter t appears in a different form.

9–2. The basic sets are given in the following tableaux:

		x_1	x_2	
	x_3	1	1	3
A	x_4	2	−1	2
	B	$-2-2u-v$	$-1-u+v$	0

Maximum if $-2u-v \geqslant 2$
and $-u+v \geqslant 1$.

		x_4	x_2	
	x_3	−1/2	3/2	2
B	x_1	1/2	−1/2	1
	B	$1+u+\tfrac{1}{2}v$	$-2-2u+\tfrac{1}{2}v$	$2+2u+v$

Maximum if $-2u-v \leqslant 2$
and $-4u+v \geqslant 4$.

		x_4	x_3	
	x_2	−1/3	2/3	4/3
C	x_1	1/3	1/3	5/3
		$\dfrac{1+u+2v}{3}$	$\dfrac{4+4u-v}{3}$	$\dfrac{14+14u+v}{3}$

Maximum if $u+2v \geqslant -1$
and $-4u+v \leqslant 4$.

		x_1	x_3	
	x_2	1	1	3
D	x_4	3	1	5
		$-1-u-2v$	$1+u-v$	$3+3u-3v$

Maximum if $u+2v \leqslant -1$
and $-u+v \leqslant 1$.

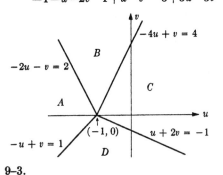

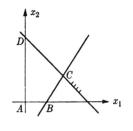

9–3.

		x_4	x_5	t over y	
−1	x_1	1/3	2/3	0	4
1	x_2	−1/3	1/3	1/3*	1
	x_3	1	1	0	9
	C	−2/3	−1/3	$1/3-t$	−3

Minimum if $t \geqslant 1/3$.

		1		
	x_4	x_5	x_2	
--1 x_1	1/3	2/3	0	4
t y	−1	1	3	3
x_3	1*	1	0	9

Minimum if $1/3 \geqslant t \geqslant -1/3$.

C	−1/3−t	−2/3+t	−1+3t	−4+3t

		1		
	x_3	x_5	x_2	
−1 x_1	−1/3	1/3	0	1
t y	1	2	3	12
x_4	1	1	0	9

Minimum if $t \leqslant -1/3$.

C	1/3+t	−1/3+2t	−1+3t	−1+12t

(From [118])

9–4. (See the figure on the following page.)

	−1+t	2+t	2+t	
	x_1	x_2	x_3	
x_4	−2	2	1	12
x_5	3*	−18	−4	24
x_6	1	2	4	24

Minimum if $t \geqslant 1$.

C	1−t	−2−t	−2−t	0

		2+t	2+t	
	x_5	x_2	x_3	
x_4	2/3	−10	−5/3	28
−1+t x_1	1/3	−6	−4/3	8
x_6	−1/3	8*	16/3	16

Minimum if $1 \geqslant t \geqslant 4/7$.

C	$\dfrac{-1+t}{3}$	4−7t	$\dfrac{-2-7t}{3}$	−8+8t

			2+t	
	x_5	x_6	x_3	
x_4	1/4	5/4	5	48
−1+t x_1	1/12	3/4	8/3	20
2+t x_2	−1/24	1/8	2/3	2

Minimum if $t \leqslant 4/7$.

C	$\dfrac{-4+t}{24}$	$\dfrac{-4+7t}{8}$	$\dfrac{-10+7t}{3}$	−16+22t

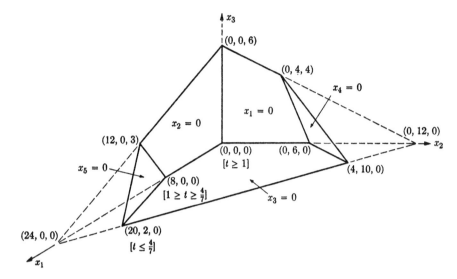

CHAPTER 10

10–1. The constraints are

$$0 \leqslant x_1 \leqslant 2,$$
$$0 \leqslant x_2 \leqslant 2,$$

Either $x_1 \leqslant 1$ or $x_2 \leqslant 1$ (or both).

The either-or constraint can be written (taking into account the other constraints as well)

$$x_1, x_2 \geqslant 0,$$
$$1 - x_1 \geqslant -d, \quad 1 - x_2 \geqslant -1 + d \quad \text{(i.e. } 2 - x_2 \geqslant d),$$
$$d = 0 \text{ or } 1. \quad \text{(From [46])}$$

10–2. (a) We first proceed as follows:

	x_3	x_4	
x_1	$-1/11$	$10/11$	$14\frac{6}{11}$
x_2	$1/11$	$1/11$	$5\frac{5}{11}$
x_5'	$-10/11$	$-10/11$*	$-\frac{6}{11}$
B	11	1	460

	x_3	x_5'	
x_1	-1	1	14
x_2	0	0.1	5.4
x_4	1	-1.1	0.6
B	10	1.1	459.4.

Introduce the new variable

$$x_6' \quad\quad 0 \quad\quad -0.1* \quad\quad -0.4.$$

One more iteration gives the final answer:

	x_3	x_6'	
x_1	-1	10	10
x_2	0	1	5
x_4	1	-11	5
x_5'	0	-10	4
B	10	11	455

Instead of x_6' we could introduce x_6'', by

$$x_6'' \quad\quad 0 \quad\quad -0.9* \quad\quad -0.6,$$

and obtain

	x_3	x_6''	
x_1	-1	$10/9$	$40/3$
x_2	0	$1/9$	$16/3$
x_4	1	$-11/9$	$4/3$
x_5'	0	$-10/9$	$2/3$
B	10	$11/9$	458
x_7'	0	$-1/9*$	$-1/3$

and finally

	x_3	x_7'	
x_1	-1	10	10
x_2	0	1	5
x_4	1	-11	5
x_5'	0	-10	3
x_6''	0	-9	3
B	10	11	455

(b) The two possibilities are the same as those which appear when we consider x_2 first.

10–3. Start from the final tableau (see Exercise 9–2):

	x_4	x_3	
x_2	$-1/3$	$2/3$	$4/3$
x_1	$1/3$	$1/3$	$5/3$
	$\dfrac{1+u+2v}{3}$	$\dfrac{4+4u-v}{3}$	$\dfrac{14+14u+v}{3}$

(a) Let $4 + 4u - v \leqslant 1 + u + 2v$, i.e. $-u + v \geqslant 1$. Then the further variable to be introduced leads to the row

$$s_1 \qquad -1/3 \qquad -1/3^* \qquad -2/3,$$

and one application of the Dual Simplex method gives

	x_4	s_1	
x_2	-1	2	0
x_1	0	1	1
x_3	1	-3	2
B	$-1 - u + v$	$4 + 4u - v$	$2 + 2u + v$

(b) Let the inequality sign in (a) be reversed; then the further row

$$s_1 \qquad -2/3^* \qquad -2/3 \qquad -1/3$$

gives

	s_1	x_3	
x_2	$-1/2$	1	$3/2$
x_1	$1/2$	0	$3/2$
x_4	$-3/2$	1	$1/2$
B	$\dfrac{1 + u + 2v}{2}$	$1 + u - v$	$\dfrac{9 + 9u}{2}$

$$s_2 \qquad -1/2^* \qquad 0 \qquad -1/2$$

	s_2	x_3	
x_2	-1	1	2
x_1	1	0	1
x_4	-3	1	2
B	$1 + u + 2v$	$1 + u - v$	$4 + 4u - v$

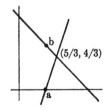

10–4. The problem can be formulated as follows:

$$2 \leqslant x_{01} + x_{12} \leqslant 3, \qquad 3 \leqslant x_{01} + x_{02},$$
$$2 \leqslant x_{02} + x_{12} \leqslant 3, \qquad 0 \leqslant x_{ij} \leqslant 2 \quad (i, j = 0, 1, 2).$$
$$\text{Minimize } 3x_{12} + 5x_{01} + 4x_{02}.$$

		5	4	3	
		x_{01}	x_{02}	x_{12}	
	y_1	−1		−1	−2
	y_2		−1	−1	−2
	z_1	1		1	3
	z_2		1	1	3
	t	−1	−1*		−3
		−5	−4	−3	0

The basic variables are slack variables. We use the bounded variable method, with the Dual Simplex algorithm.

		5		3	
		x_{01}	t	x_{12}	
	y_1	−1	0	−1	−2
	y_2	1	−1	−1	1
	z_1	1	0	1	3
	z_2	−1	1	1	0
4	x_{02}	1	−1	0	3
		−1	−4	−3	12

x_{02} exceeds its upper limit, and therefore we introduce its complement, x'_{02}, say.

					(−8)
	y_1	−1*	0	−1	−2
	y_2	1	−1	−1	1
	z_1	1	0	1	3
	z_2	−1	1	1	0
−4	x'_{02}	−1	1	0	−1
		−1	−4	−3	12

This corresponds to the constant term in the expression of the objection function.

		y_1	t	x_{12}	
					(−8)
5	x_{01}	−1	0	1	2
	y_2	1	−1	−2*	−1
	z_1	1	0	0	1
	z_2	−1	1	2	2
−4	x'_{02}	−1	1	1	1
		−1	−4	−2	14

		y_1	t	y_2	(-8)
5	x_{01}	$-\frac{1}{2}$	$-\frac{1}{2}$	$\frac{1}{2}$	$\frac{3}{2}$
3	x_{12}	$-\frac{1}{2}$	$\frac{1}{2}$	$-\frac{1}{2}$	$\frac{1}{2}$
	z_1	1	0	0	1
	z_2	0	0	1	1
-4	x_{02}'	$-\frac{1}{2}$	$\frac{1}{2}$	$\frac{1}{2}$	$\frac{1}{2}$
		-2	-3	-1	15
	s	$-\frac{1}{2}$	$-\frac{1}{2}$	$-\frac{1}{2}*$	$-\frac{1}{2}$

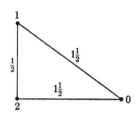

End of the first part.

		y_1	t	s	(-8)
5	x_{01}	-1	-1	1	1
3	x_{12}	0	1	-1	1
	z_1	1	0	0	1
	z_2	-1	-1	2	0
-4	x_{02}'	-1	0	1	0
	y_2	1	1	-2	1
		-1	-2	-2	16

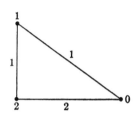

10–5. The solution is obtained from:

		0.06			
		x_1	y_2	y_3	
	y_1	1	0	0	15
0.15	x_2	0.3	1	0	9
0.3	x_3	0.09	-0.3	1	2.7
	n_2	0.3	0	0	9
	n_3	0.09	-0.3	0	2.7
	s_1	-0.09	$-0.7*$	0	-0.7
	B	0.012	0.06	0.3	16.86

		0.06			
		x_1	s_1	y_3	
	y_1	1	0	0	15
0.15	x_2	12/70	10/7	0	8
0.3	x_3	9/70	$-3/7$	1	3
	n_2	3/10	0	0	9
	n_3	9/70	$-3/7$	0	3
	y_2	9/70	$-10/7$	0	1
	B	3/700	6/70	3/10	16.8

10-6. The optimum for continuous variables is (see Examples 5-1 and 5-7)

		x_4	x_5	2 x_3	
3	x_1	3/5	−4/5	−1	2/5
6	x_2	−1/5	3/5	1	1/5
	B	3/5	6/5	1	12/5

		x_4	x_5	x_3	t_1	
	x_2	−1/5	3/5	1	0	1/5
$H_1 = B + t_1 - 12/5$		3/5*	6/5	1	−1	0
	B	3/5	−4/5	−1	0	2/5

	H_1	x_5	x_3	t_1	
x_2	1/3	1	4/3	−1/3	1/5
x_4	5/3	2	5/3	−5/3	0
x_1	−1	−2	−2	1	2/5

We want to determine t_1 such that $x_1 + t_1 = 2/5$, $x_1 = 0$; hence $t_1 = 2/5$, and therefore $x_2 = 1/3$, $x_4 = 2/3$, $B + 2/5 - 12/5 = 0$, i.e. $B = 2$.

	x_3	x_5	$B - 2$	
x_1	−2	−2	−1	0
x_4	5/3	2	5/3	2/3
F	−2	−2	−1*	0
x_2	4/3	1	1/3	1/3

	x_3	x_5	F	t_2	
x_1	0	0	−1	0	0
x_4	−5/3	−4/3	5/3	0	2/3
$H_2 = B - 2 + t_2$	2*	2	−1	−1	0
x_2	2/3	1/3	1/3	0	1/3

	H_2	x_5	F	t_2	
x_1	0	0	−1	0	0
x_4	5/6	1/3	5/6	−5/6	2/3
x_3	1/2	1	−1/2	−1/2	0
x_2	−1/3	−1/3	2/3	1/3	1/3

To make $x_2 = 0$, we have $t_2 = 1$, and hence

$$x_1 = x_2 = 0, \qquad x_3 = 1/2, \qquad x_4 = 3/2.$$

It follows from the constraints that $x_1 = 1$ is not feasible. When $x_1 = 0$, then x_2 is not feasible, so we have reached the complete final answer.

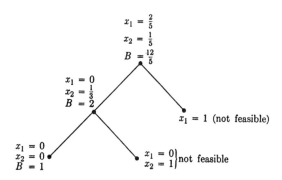

CHAPTER 11

11–1. The sets of basic variables are (x_1, x_2) or (x_1, x_3) or (x_2, x_3). We have, respectively,

$$x_1 = \frac{1 + 2b_{1t} - b_{2t} + x_3}{3},$$

$$x_2 = \frac{1 - b_{1t} + 2b_{2t} + x_3}{3}.$$

Minimize $\dfrac{2}{3} + \dfrac{b_{1t} + b_{2t}}{3} + \dfrac{5x_3}{3}$;

or

$$x_1 = b_{1t} - b_{2t} + x_2,$$
$$x_3 = -1 + b_{1t} - 2b_{2t} + 3x_2.$$

Minimize $-1 + 2b_{1t} - 3b_{2t} + 5x_2$;

or

$$x_2 = -b_{1t} + b_{2t} + x_1,$$
$$x_3 = -1 - 2b_{1t} + b_{2t} + 3x_1.$$

Minimize $-1 - 3b_{1t} + 2b_{2t} + 5x_1.$

Hence the following are the optimal solutions for the various pairs of values of b_{1t} and b_{2t}.

b_{1t}	b_{2t}	x_1	x_2	x_3	Objective function
0	0	$\frac{1}{3}$	$\frac{1}{3}$	0	$\frac{2}{3}$
2	0	2	0	1	3
0	2	0	2	1	3
2	2	1	1	0	2
−2	0	0	2	3	5
0	−2	2	0	3	5
−2	−2	0	0	1	1
2	−2	4	0	5	9
−2	2	0	4	5	9

The value of the objective function for $b_{1t} = b_{2t} = 0$ is clearly smaller than the average of the other values, whatever the p_t.

11–2. (a) We have

$$\min_t b_{1t} = \min_t b_{2t} = -2.$$

The optimal solution for

$$2x_1 + x_2 - x_3 \leqslant -1,$$
$$x_1 + 2x_2 - x_3 \leqslant -1,$$
$$\text{Minimize } 2 - 2x_1 - 2x_2 + 3x_3,$$

is

$$x_1 = x_2 = 0, \qquad x_3 = 1,$$
$$\text{Objective function} = 5 \quad (> M_0 = 2/3).$$

(b) We have

$$2x_1 + x_2 - x_3 \leqslant 1,$$
$$x_1 + 2x_2 - x_3 \leqslant 1,$$
$$\text{Minimize } 2 - 2x_1 - 2x_2 + 3x_3.$$

The optimal solution is

$$x_1 = x_2 = \tfrac{1}{3}, \qquad x_3 = 0,$$
$$\text{Objective function} = \tfrac{2}{3} \quad (< M_1 = 5, = M_0 = \tfrac{2}{3}).$$

CHAPTER 12

12–1. We write the constraints as

$$t_0 = x_0 - 3, \qquad t_1 = 6 - x_1,$$
$$x_0, x_1 \geqslant 0.$$

Note that t_0 and t_1 are not sign-restricted. They can be treated as "free variables." Thus

$$
\begin{array}{ll}
180 + 2t_0 - 28t_1 & x_0 = 3 + t_0 \\
+ (\quad 2 + 4t_0 + 2t_1)t_0 & x_1 = 6 - t_1. \\
+ (-28 + 2t_0 + 6t_1)t_1 &
\end{array}
$$

Start with $t_0 = t_1 = 0$. It is useful to increase t_1, but not beyond $14/3$. Introduce $u_1 = -28 + 2t_0 + 6t_1$. This gives

$$
\begin{array}{ll}
148/3 + 34t_0/3 & x_0 = 3 + t_0 \\
+ (\ 34/3 + 10t_0/3 \quad)t_0 & t_1 = 14/3 - t_0/3 + u_1/6. \\
+ (\qquad\qquad u_1/6)u_1 & x_1 = 4/3 + t_0/3 - u_1/6
\end{array}
$$

It is useful to decrease t_0, but not below -3. This gives

$$
\begin{array}{ll}
34/3 + 4x_0/3 & t_0 = -3 + x_0 \\
+ (\ 4/3 + 10x_0/3 \quad)x_0 & t_1 = 17/3 - x_0/3 + u_1/6. \\
+ (\qquad\qquad u_1/6)u_1 & x_1 = 1/3 + x_0/3 - u_1/6
\end{array}
$$

Thus the minimum is $34/3$, obtained when $t_0 = -3$, $t_1 = 17/3$. Compare this with the result of Section 8–4, where the t_i were unrestricted.

A second way of solving this problem starts with eliminating t_0 and t_1, and introducing x_0 and x_1 (sign-restricted) instead. This gives

$$
4x_0^2 - 4x_0x_1 + 6x_1^2 + 4x_0 - 4x_1 + 12.
$$

Thus

$$
\begin{array}{l}
12 + 2x_0 - 2x_1 \\
+ (\quad 2 + 4x_0 - 2x_1)x_0 \\
+ (-2 - 2x_0 + 6x_1)x_1.
\end{array}
$$

Introduce $u_1 = -2 - 2x_0 + 6x_1$, i.e. $x_1 = 1/3 + x_0/3 + u_1/6$:

$$
\begin{array}{l}
34/3 + 4x_0/3 \\
+ (\ 4/3 + 10x_0/3 \qquad)x_0 \\
+ (\qquad\qquad u_1/6)u_1.
\end{array}
$$

This is already the final stage. Returning to t_0 and t_1, we have the same result as before.

12–2. We have

$$
\begin{array}{ll}
5 - 2x_1 - x_2 & \\
+ (-2 + x_1 \quad)x_1 & x_1 + x_2 + x_3 = 4. \\
+ (-1 \qquad + x_2)x_2, &
\end{array}
$$

Start with $x_1 = x_2 = 0$. Increase x_1; hence introduce $u_1 = -2 + x_1$:

$$1 \qquad - x_2 \qquad x_1 = 2 + u_1,$$
$$+ (\qquad u_1 \qquad)u_1 \qquad x_3 = 2 - u_1 - x_2.$$
$$+ (-1 \qquad + x_2)x_2,$$

Increase x_2; hence introduce $u_2 = -1 + x_2$. The objective function becomes $u_1^2 + u_2^2$, and the constraints are now

$$x_1 = 2 + u_1, \qquad x_2 = 1 + u_2, \qquad x_3 = 1 - 2u_1.$$

The minimum is obtained at $x_1 = 2$, $x_2 = 1$, which is the minimum obtained by

$$\frac{\partial C}{\partial x_1} = \frac{\partial C}{\partial x_2} = 0.$$

12–3. Form (a) equals

$$(x_1 + x_2 + x_3)^2 + (2x_1 - x_2 - x_3)^2,$$

and is therefore positive semidefinite. It is zero, for instance, for $x_1 = 0$, $x_2 = 1$, $x_3 = -1$.

Form (b) equals

$$(x_1 + x_2 + x_3)^2 + (2x_1 - x_2 - x_3)^2 + x_2^2,$$

and is therefore positive definite. It is only zero for $x_1 = x_2 = x_3 = 0$.

Form (c) equals

$$(x_1 + x_2 + x_3)^2 + (2x_1 - x_2 - x_3)^2 - x_2^2,$$

and is therefore neither. It is, for instance, zero for $x_1 = 1$, $x_2 = 1$, $x_3 = -2$, and negative for $x_1 = 0$, $x_2 = 1$, $x_3 = -1$.

12–4. The dual problem is

Maximize $183 - 8u_1^2 + 12u_1u_2 - 17u_2^2 - 10y_1$,
subject to
$$2y_1 + 16u_1 - 12u_2 \geqslant 44,$$
$$y_1 - 12u_1 + 34u_2 \geqslant 42,$$
$$y_1 \geqslant 0.$$

We introduce slack variables, say z_1 and z_2, and solve the constraints for u_1 and u_2; thus

$$u_1 = 5 - 0.2y_1 + 0.085z_1 + 0.03z_2,$$
$$u_2 = 3 - 0.1y_1 + 0.03z_1 + 0.04z_2.$$

Substituting into the objective function and into the constraints we obtain

$$10 \quad + 1.5y_1 \quad - 2.5z_1 \quad - 0.9z_2$$
$$+ (\quad 1.5 - 0.25y_1 \quad + 0.1z_1 \quad + 0.026z_2)y_1$$
$$+ (-2.5 + 0.1y_1 \quad - 0.02125z_1 - 0.015z_2)z_1$$
$$+ (-0.9 + 0.026y_1 - 0.015z_1 \quad - 0.01z_2)z_2,$$

subject to $y_1, z_1, z_2 \geqslant 0$.

It is useful to increase y_1, but not beyond $1.5/0.25 = 6$. This gives

$$y_1 = 6, \qquad u_1 = 3.8, \qquad u_2 = 2.4,$$
$$\text{Objective function} = 19.$$

CHAPTER 13

13-1. We have

$$f_1(n) = \max_{0 \leqslant x \leqslant n} [3x + 2.5(n-x)] = 3n \qquad (x = n),$$
$$f_2(n) = \max_{0 \leqslant x \leqslant n} [3x + 2.5(n-x) + (2n - x)] = 4.5n \qquad (x = 0),$$
$$f_3(n) = \max_{0 \leqslant x \leqslant n} \{3x + 2.5(n-x) + 4.5[\tfrac{1}{3}x + \tfrac{2}{3}(n-x)]\} = 5.5n \qquad (x = 0).$$

Schedule

Stage	First job	Second job	Result	Remaining
1	0	n	$5n/2$	$2n/3$
2	0	$2n/3$	$5n/3$	$4n/9$
3	$4n/9$	0	$4n/3$	
			$5.5n$	

13-2. After the first month, the following numbers of items can be left over, and the following decisions can be made:

Decisions

	Return								Buy	
Left-over	7	6	5	4	3	2	1	0	1	2
0								0	$9\frac{1}{4}$	7
1							5	$19\frac{1}{4}$	15	
2						10	$24\frac{1}{4}$	27		
3					15	$29\frac{1}{4}$	32			
4				20	$34\frac{1}{4}$	37				
5			25	$39\frac{1}{4}$	42					
6		30	$44\frac{1}{4}$	47						
7	35	$49\frac{1}{4}$	52							

(More than 7 will not be bought to begin with, since they will not be required during the two months.)

The entries in the table indicate the expected profit, given the number of left-overs, and the decision made. For instance, if 4 are left, and 2 of them returned, then 10 is obtained at once, and

> 4 more if none of those remaining are sold,
> 27 more if one more is sold and one returned,
> 50 more if both remaining are sold.

The expected value of the income is then $14/4 + 37/2 + 60/4 = 37$.

Clearly, whatever the number of left-overs after the first month, that decision is to be made which produces, according to the table above, the highest income.

We turn to the first month. The following table contains the expected income, dependent on the decision of how many items to buy, and the actual experienced requirement:

Required in the first month

Bought	0	1	2	3	4	5	Expected value
0	$9\frac{1}{4}$	$9\frac{1}{4}$	$9\frac{1}{4}$	$9\frac{1}{4}$	$9\frac{1}{4}$	$9\frac{1}{4}$	9.25
1	$9\frac{1}{4}$	$24\frac{1}{4}$	$24\frac{1}{4}$	$24\frac{1}{4}$	$24\frac{1}{4}$	$24\frac{1}{4}$	22.375
2	7	$24\frac{1}{4}$	$39\frac{1}{4}$	$39\frac{1}{4}$	$39\frac{1}{4}$	$39\frac{1}{4}$	33.34375
3	2	22	$39\frac{1}{4}$	$54\frac{1}{4}$	$54\frac{1}{4}$	$54\frac{1}{4}$	39.9375
4	−3	17	37	$54\frac{1}{4}$	$69\frac{1}{4}$	$69\frac{1}{4}$	41.875 (max)
5	−8	12	32	52	$69\frac{1}{4}$	$84\frac{1}{4}$	40.6875
6	−13	7	27	47	67	$84\frac{1}{4}$	36.65625
7	−18	2	22	42	62	82	32

For instance, if there were 4 items bought and 2 sold, then there was an income of $50 - 40$ in the first month, and the two items left over will, according to the first table, produce 27 more, when the best decision is made, viz. not to buy anymore but not to return any either. Thus the entry 27 above is obtained. It follows, that at the start 4 items should be bought, since the expected value 41.875 is larger than any other.

13–3. When we are in Y, S, or D, then there is only one single direct path into H, and this is shorter than any roundabout path. From L, we could go to Y (L–Y–H has length 62), to S (L–S–H = 55), to D (L–D–H = 75), to B (L–B–D–H = 83), or to M (certainly too long to get to H). Hence we would go to S. From B, we have the choices to L (B–L, then optimal to H, has length 74), to D (B–D–H = 62), or to M (certainly too long). Hence we would go to D and H. From M, we can go to L (M–L optimal to H = 95), or to B (M–B optimal to H = 98). Hence we choose M–L–S–H.

13-4. Denote the minimum deviation when the optimal directions are used throughout at and nearer than a distance x from the opposite bank by $F(x)$. If the angle between the line orthogonal to the river's flow and the

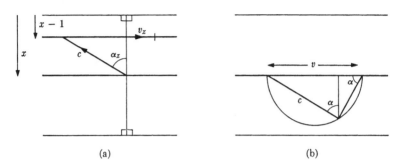

(a)   (b)

direction of swimming is α_x [see diagram (a)], then the deviation at a distance $x - 1$ will have increased from that at x by

$$\frac{v}{c \times \cos \alpha_x} - \tan \alpha_x.$$

Hence

$$F(x) = \min \left[\frac{v}{c \times \cos \alpha_x} - \tan \alpha_x + F(x - 1) \right].$$

If $v_x = v$, then

$$g(x) = \frac{xv}{c \times \cos \alpha} - x \tan \alpha$$

is a solution of

$$g(x) = \frac{v}{c \times \cos \alpha} - \tan \alpha + g(x - 1)$$

for any value of α and hence the required direction is given by

$$\frac{d}{d\alpha} \left(\frac{v}{c \times \cos \alpha} - \tan \alpha \right) = 0,$$

i.e. $\alpha = \sin^{-1} c/v$. Diagram (b) indicates a convenient way of finding this α.

BIBLIOGRAPHY

This bibliography contains only a selection of titles of books and papers on the subjects of this book. In May 1958 there appeared an excellent annotated bibliography, containing all relevant items up to June 1957, and a good many more. The precise reference is: Vera Riley and Saul I. Gass, *Linear Programming and Associated Techniques: A Comprehensive Bibliography on Linear, Nonlinear, and Dynamic Programming*, Operations Research Office, Johns Hopkins University, Chevy Chase, Maryland, May 1958.
In the present bibliography we mention a few books, some of them referred to in the text, and papers and reports, all of which are referred to.

BOOKS

1. ANTOSIEWICZ, H. (edit.), *Second Symposium in Linear Programming*, Washington, 1955.
2. ARROW, K. J., L. HURWITZ, and H. UZAWA, "Studies in Linear and Nonlinear Programming," *Stanford Mathematical Studies in the Social Sciences II*, Stanford, Calif., 1958.
3. BELLMAN, RICHARD, *Dynamic Programming*, Princeton, N.J., 1957.
4. BERGE, C., *Théorie des graphes et ses applications*, Paris, 1958.
5. CHARNES, A., W. W. COOPER, and A. HENDERSON, *An Introduction to Linear Programming*, New York, N.Y., 1953.
6. DORFMAN, R., P. A. SAMUELSON, and R. M. SOLOW, *Linear Programming and Economic Analysis*, New York, N.Y., 1958.
7. GASS, SAUL I., *Linear Programming: Methods and Applications*, New York, N.Y., 1958.
8. KARLIN, SAMUEL, *Mathematical Methods and Theory in Games, Programming, and Economics*, Reading, Mass., 1959.
9. KÖNIG, D., *Theorie der endlichen und unendlichen Graphen*, Leipzig, 1936, New York, N.Y., 1950.
10. KOOPMANS, TJALLING C. (edit.), *Activity Analysis of Production and Allocation*, Monograph No. 13 of the Cowles Commission, New York, N.Y., 1951.
11. KUHN, HAROLD W., and ALBERT W. TUCKER (edit.), "Linear Inequalities and Related Systems," *Annals of Mathematics Studies No. 38*, Princeton, N.J., 1956.
12. LEONTIEF, W. W., *The Structure of the American Economy, 1919–1939: An Empirical Application of Equilibrium Analysis*, New York, 1951.
13. METZGER, R. W., *Elementary Mathematical Programming*, New York, N.Y., 1958.
14. VAJDA, S., *The Theory of Games and Linear Programming*, London, 1956.
15. ——, *Readings in Linear Programming*, London, 1958.

16. ——, *An Introduction to Linear Programming and the Theory of Games*, London, 1960.

17. WOLFE, P. (edit.), *The RAND Symposium on Mathematical Programming: Linear Programming and Recent Extensions*, Proceedings of a Conference held March 16–20, 1959, The RAND Corporation, Santa Monica, Calif., 1960.

PAPERS AND REPORTS

Abbreviations

Bull. Am. Math. Soc.	*Bulletin of the American Mathematical Society*
Can. J. Math.	*Canadian Journal of Mathematics*
Econ.	*Econometrica*
J.A.S.A.	*Journal of the American Statistical Association*
J.O.R.S.A.	*Journal of the Operations Research Society of America*, later renamed *Operations Research*
J. Roy. Stat. Soc.	*Journal of the Royal Statistical Society.*
Man. Sci.	*Management Science*
Nav. Res. Log. Qu.	*Naval Research Logistics Quarterly*
Op. Res. Qu.	*Operational Research Quarterly*
Pac. J. Math.	*Pacific Journal of Mathematics*

18. ANTOSIEWICZ, HENRY A., "A Theorem on Alternatives for Pairs of Matrices," *Pac. J. Math.*, 1955, Suppl. 1, **5**, 641–642.

19. ——, and ALAN J. HOFFMAN, "A Remark on the Smoothing Problem," *Man. Sci.*, 1954, **1**, 92–95.

20. BABBAR, M. M., "Distributions of Solutions of a Set of Linear Equations (with an application to Linear Programming)," *J.A.S.A.*, 1955, **50**, 854–869.

21. BARANKIN, EDWARD W., and ROBERT DORFMAN, "On Quadratic Programming," *University of California Publications in Statistics*, 1958, **2**, 285–317.

22. BARNA, TIBOR, "The Interdependence of the British Economy," *J. Roy. Stat. Soc.* (A), 1952, **115**, 29–77.

23. BEALE, E. M. L., "An Alternative Method for Linear Programming," *Proceedings of the Cambridge Philosophical Society*, 1954, **50**, 513–523.

24. ——, "On Optimizing a Convex Function Subject to Linear Inequalities," *J. Roy. Stat. Soc.*, 1955, **17**, 173–184.

25. ——, "Cycling in the Dual Simplex Algorithm," *Nav. Res. Log. Qu.*, 1955, **2**, 269–275.

26. ——, "An Algorithm for the Warehouse Problem," Technical Report No. 22, Statistical Techniques Research Group, Princeton University, 1958.

27. ——, "An Algorithm for Solving the Transportation Problem when the Shipping Cost over each Route is Convex," *Nav. Res. Log. Qu.*, 1959, **6**, 43–56.

28. ——, "On Quadratic Programming," *Nav. Res. Log. Qu.*, 1959, **6**, 227–243.

29. ——, G. MORTON, and A. H. LAND, "Solution of a Purchase Storage Programme," *Op. Res. Qu.*, 1958, **9**, 174–197.

30. BELLMAN, RICHARD E., "Equipment Replacement Policy," *Journal of the Society for Industrial and Applied Mathematics*, 1955, **3**, 133–136.

31. ——, "Some Aspects of the Theory of Dynamic Programming," *Scripta Mathematica*, 1955, **21**, 273–277.

32. ——, "On a Dynamic Programming Approach to the Caterer Problem: —I," *Man. Sci.*, 1957, **3**, 270–278.

33. CAHN, ALBERT S., Jr., "The Warehouse Problem," Abstract, *Bull. Am. Math. Soc.*, 1948, **54**, 1073.

34. CHARNES, ABRAHAM, "Optimality and Degeneracy in Linear Programming," *Econ.*, 1952, **20**, 160–170.

35. ——, and WILLIAM COOPER, "Management Models and Industrial Applications of Linear Programming," *Man. Sci.*, 1957, **4**, 38–91.

36. ——, ——, and BOB MELLON, "A Model for Programming and Sensitivity Analysis in an Integrated Oil Company," *Econ.*, 1954, **22**, 193–217.

37. ——, and CARLTON E. LEMKE, "Minimization of Non-linear Separable Convex Functionals," *Nav. Res. Log. Qu.*, 1954, **1**, 301–312.

38. DANTZIG, GEORGE B., "Programming of Interdependent Activities: II, Mathematical Model," Chapter II in [10], 1951, and also *Econ.*, 1949, **17**, 200–211.

39. ——, "Maximization of a Linear Function of Variables Subject to Linear Inequalities," Chapter XXI in [10], 1951.

40. ——, "Application of the Simplex Method to a Transportation Problem," Chapter XXIII in [10], 1951.

41. ——, "Upper Bounds, Secondary Constraints, and Block Triangularity in Linear Programming," *Econ.*, 1955, **23**, 174–183.

42. ——, "Optimal Solution of a Dynamic Leontief Model with Substitution," *Econ.*, 1955, **23**, 295–302.

43. ——, "Linear Programming under Uncertainty," *Man. Sci.*, 1955, **1**, 197–206.

44. ——, "Note on B. Klein's Direct Use of Extremal Principles in Solving Certain Optimizing Problems Involving Inequalities," *J.O.R.S.A.*, 1956, **4**, 247–249.

45. ——, "Note on Solving Linear Programs in Integers," *Nav. Res. Log. Qu.*, 1959, **6**, 75–76.

46. ——, "On the Significance of Solving Linear Programs with Some Integer Variables," *Econ.*, 1960, **28**, 30–44.

47. ——, "On the Shortest Route Through a Network," *Man. Sci.*, 1960, **6**, 187–190.

48. ——, "A Machine-Job Scheduling Model," *Man. Sci.*, 1960, **6**, 191–196.

49. ——, LESTER R. FORD, Jr., and DELBERT R. FULKERSON, "A Primal-Dual Algorithm for Linear Programs," Paper 7 in [11], 1956.

50. ——, and DELBERT R. FULKERSON, "On the Max-Flow Min-Cut Theorem of Networks," Paper 12 in [11], 1956.

51. ——, ——, and SELMER JOHNSON, "Solution of a Large-Scale Traveling Salesman Problem," *J.O.R.S.A.*, 1954, **2**, 393–410.

52. ——, ——, and ——, "On a Linear-Programming Combinatorial Approach to the Traveling Salesman Problem," *J.O.R.S.A.*, 1959, **7**, 58–66.

53. ——, SELMER JOHNSON, and WAYNE WHITE, "A Linear Programming Approach to the Chemical Equilibrium Problem," *Man. Sci.*, 1958, **5**, 38–43.

54. ——, and WILLIAM ORCHARD-HAYS, "The Product Form for the Inverse in the Simplex Method," *Mathematical Tables and Other Aids to Computation*, 1954, **8**, 64–67.

55. ——, ALEX ORDEN, and PHILIP WOLFE, "The Generalized Simplex Method for Minimizing a Linear Form under Linear Inequality Constraints," *Pac. J. Math.*, 1955, **5**, 183–195.

56. DENNIS, JACK RONNELL, "Mathematical Programming and Electrical Networks," Internal Technical Report No. 10, Massachusetts Institute of Technology, 1958.

57. DOIG, ALISON, and AILSA H. LAND, "An Automatic Method of Solving Discrete Programming Problems," *Econ.*, 1960, **28**, 497–520.

58. DORN, WILLIAM S., "Duality in Quadratic Programming," Report NY08676 (Physics), New York University, 1958.

59. DREYFUS, STUART E., "An Analytic Solution of the Warehouse Problem," *Man. Sci.*, 1957, **4**, 99–104.

60. EGERVÁRY, E., "On Combinatorial Properties of Matrices," Logistics Papers No. 4, George Washington University Research Project, 1953 (Translation by Harold W. Kuhn).

61. EISEMANN, KURT, "The Trim Problem," *Man. Sci.*, 1957, **3**, 279–284.

62. FARKAS, J., "Über die Theorie der einfachen Ungleichungen," *Journal für die reine und angewandte Mathematik*, 1902, **124**, 1–24.

63. FERGUSON, ALLEN R., and GEORGE B. DANTZIG, "The Allocation of Aircraft to Routes—An Example of Linear Programming under Uncertain Demand," *Man. Sci.*, 1956, **3**, 45–73.

64. FLOOD, MERRILL M., "The Traveling Salesman Problem," *J.O.R.S.A.*, 1956, **4**, 61–75.

65. FORD, LESTER R., Jr., and DELBERT R. FULKERSON, "A Simple Algorithm for Finding Maximal Network Flows and an Application to the Hitchcock Problem," *Can. J. Math.*, 1957, **9**, 210–218.

66. ——, ——, "Solving the Transportation Problem," *Man. Sci.*, 1956, **3**, 24–32.

67. ——, ——, "A Primal-Dual Algorithm for the Capacitated Hitchcock Problem," *Nav. Res. Log. Qu.*, 1957, **4**, 47–54.

68. FRANK, MARGUERITE, and PHILIP WOLFE, "An Algorithm for Quadratic Programming," *Nav. Res. Log. Qu.*, 1956, **3**, 95–110.

69. FRISCH, RAGNAR A. K., "The Multiplex Method for Linear Programming," *Sankhya*, 1957, **18**, 329–362.

70. GALE, DAVID, "The Basic Theorems of Real Linear Equations, Inequalities, Linear Programming, and Game Theory," *Nav. Res. Log. Qu.*, 1956, **3**, 193–200.

71. ——, "A Theorem on Flows in Networks," *Pac. J. Math.*, 1957, **7**, 1073–1082.

72. ——, HAROLD W. KUHN, and ALBERT W. TUCKER, "Linear Programming and the Theory of Games," Chapter XIX in [10], 1951.

73. GASS, SAUL I., and THOMAS SAATY, "The Computational Algorithm for the Parametric Objective Function," *Nav. Res. Log. Qu.*, 1955, **2**, 39–45.

74. GOLDMAN, ALAN J., and ALBERT W. TUCKER, "Theory of Linear Programming," Paper 4 in [11], 1956.

75. GOMORY, RALPH E., "Outline of an Algorithm for Integer Solutions to Linear Programs," *Bull. Am. Math. Soc.*, 1958, **64**, 275–278.

76. GOOD, R. A., "Systems of Linear Relations," *Review of the Society for Industrial and Applied Mathematics*, 1959, **1**, 1–31.

77. GORDAN, P., "Über die Lösungen linearer Gleichungen mit reellen Koeffizienten," *Math. Ann.*, 1873, **6**, 23–28.

78. HADLEY, GEORGE F., and MICHAEL A. SIMONNARD, "A Simplified Two-Phase Technique for the Simplex Method," *Nav. Res. Log. Qu.*, 1959, **6**, 221–226.

79. HALL, Jr., MARSHALL, "A Survey of Combinatorial Analysis," *Surveys in Applied Mathematics*, Vol. 4: *Some Aspects of Analysis and Probability*, New York, N.Y., 1958.

80. HALL, PHILIP, "On Representations of Subsets," *Journal of the London Mathematical Society*, 1935, **10**, 26–30.

81. HARTLEY, H. O., *Nonlinear Programming for Separable Objective Functions and Constraints*, To appear.

82. HELLER, ISIDOR, and CHARLES S. TOMPKINS, "An Extension of a Theorem of Dantzig's," Paper 14 in [11], 1956.

83. HILDRETH, CLIFFORD, "A Quadratic Programming Procedure," *Nav. Res. Log. Qu.*, 1957, **4**, 79–85.

84. HIRSCH, WARREN M., and GEORGE B. DANTZIG, "The Fixed Charge Problem," The RAND Corporation, RM-1383, 1954.

85. HITCHCOCK, FRANK L., "The Distribution of a Product from Several Sources to Numerous Localities," *Journal of Mathematics and Physics*, 1941, **20**, 224–230.

86. HOFFMAN, ALAN J., and WALTER JACOBS, "Smooth Patterns of Production," *Man. Sci.*, 1954, **1**, 86–91.

87. ——, and JOSEPH B. KRUSKAL, "Integral Boundary Points of Convex Polyhedra," Paper 13 in [11], 1956.

88. HOUTHAKKER, H. S., "The Capacity Method of Quadratic Programming," *Econ.*, 1960, **28**, 62–87.

89. JACOBS, WALTER W., "The Caterer Problem," *Nav. Res. Log. Qu.*, 1954, **1**, 154–165.

90. JOHNSON, SELMER M., "Optimal Two- and Three-Stage Production Schedules with Setup Times Included," *Nav. Res. Log. Qu.*, 1954, **1**, 61–68.

91. ——, "Sequential Production Planning over Time at Minimum Cost," *Man. Sci.*, 1957, **3**, 435–437.

92. KELLEY, JAMES E., Jr., "An Application of Linear Programming to

Curve Fitting," *Journal of the Society of Industrial and Applied Mathematics*, 1958, **6**, 15–22.

93. KLEIN, BERTRAM, "Direct Use of Extremal Principles in Solving Certain Optimizing Problems Involving Inequalities," *J.O.R.S.A.*, 1955, **3**, 168–175.

94. KOOPMANS, TJALLING C., "Optimum Utilization of the Transportation System," *Econ.*, 1949, **17** (Suppl.), 136–145.

95. ——, and STANLEY REITER, "A Model of Transportation," Chapter XIV of [10], 1951.

96. KUHN, HAROLD W., "The Hungarian Method for the Assignment Problem," *Nav. Res. Log. Qu.*, 1955, **2**, 83–97.

97. ——, "Variants of the Hungarian Method for the Assignment Problem," *Nav. Res. Log. Qu.*, 1956, **3**, 253–258.

98. ——, and ALBERT W. TUCKER, "Non-Linear Programming," Proceedings of the Second Berkeley Symposium on Mathematical Statistics and Probability (edit., J. Neyman), 1950, 481–492.

99. LEMKE, CARLTON E., "The Dual Method of Solving the Linear Programming Problem," *Nav. Res. Log. Qu.*, 1954, **1**, 36–47.

100. MADANSKY, ALBERT, "Bounds on the Expectation of a Convex Function of a Multivariate Random Variable," *Annals of Mathematical Statistics*, 1959, **30**, 743–746.

101. ——, "Inequalities for Stochastic Linear Programming Problems," *Man. Sci.*, 1960, **6**, 197–204.

102. MANNE, ALAN S., "A Note on the Modigliani-Hohn Production Smoothing Model," *Man. Sci.*, 1957, **3**, 371–379.

103. MARKOWITZ, HARRY M., "The Optimization of a Quadratic Function Subject to Linear Constraints," *Nav. Res. Log. Qu.*, 1956, **3**, 111–133.

104. ——, and ALAN S. MANNE, "On the Solution of Discrete Programming Problems," *Econ.*, 1957, **25**, 84–110.

105. MOTZKIN, THEODORE S., "Beiträge zur Theorie der linearen Ungleichungen," Dissertation Basel, 1933, Jerusalem, 1936.

106. ——, and ISAAC J. SCHOENBERG, "The Relaxation Method for Linear Inequalities," *Can. J. Math.*, 1954, **6**, 393–404.

107. ORDEN, ALEX, "A Procedure for Handling Degeneracy in the Transportation Problem," Internal U.S.A.F. Report, 1951.

108. ——, "The Transshipment Problem," *Man. Sci.*, 1956, **2**, 276–285.

109. PEARSON, CARL E., "Note on Linear Programming," *Quarterly of Applied Mathematics*, 1956, **14**, 205–206.

110. SAATY, THOMAS, and SAUL I. GASS, "Parametric Objective Function: Part I," *J.O.R.S.A.*, 1954, **2**, 316–319.

111. SCHELL, EMIL D., "Distribution of a Product by Several Properties," In [1], 615–642, 1956.

112. SIMONNARD, MICHAEL A., and GEORGE F. HADLEY, "Maximum Number of Iterations in the Transportation Problem," *Nav. Res. Log. Qu.*, 1959, **6**, 125–129.

113. STIEMKE, E., "Über positive Lösungen homogener linearer Gleichungen," *Math. Ann.*, 1915, **76**, 340–342.

114. STIGLER, GEORGE J., "The Cost of Subsistence," *Journal of Farm Economics*, 1945, **27**, 303–314.

115. SUZUKI, YUKIO, "Note on Linear Programming," *Annals of the Institute of Statistical Mathematics*, 1959, **10**, 89–105.

116. TINTNER, GERHARD, "Stochastic Linear Programming with Applications to Agricultural Economics," In [1], 197–228, 1956.

117. TUCKER, ALBERT W., "Dual Systems in Homogeneous Linear Relations," Paper 1 in [11], 1956.

118. VAJDA, STEVEN, "Inequalities in Stochastic Linear Programming," *Bulletin of the International Statistical Institute*, 1958, **36**, 357–363.

119. VAZSONYI, ANDREW, "Optimizing a Function of Additively Separated Variables Subject to a Simple Restriction," In [1], 453–469, 1956.

120. WAGNER, HARVEY M., "A Two-Phase Method for the Simplex Tableau," *J.O.R.S.A.*, 1956, **4**, 443–447.

121. ——, "A Comparison of the Original and the Revised Simplex Methods," *J.O.R.S.A.*, 1957, **5**, 361–369.

122. ——, "The Dual Simplex Algorithm for Bounded Variables," *Nav. Res. Log. Qu.*, 1958, **5**, 257–261.

123. ——, "On the Distribution of Solutions in Linear Programming Problems," *J.A.S.A.*, 1958, **53**, 161–163.

124. ——, "An Integer Linear Programming Model for Machine Scheduling," *Nav. Res. Log. Qu.*, 1959, **6**, 131–140.

125. ——, "On the Capacitated Hitchcock Problem," Technical Report No. 54, Department of Economics, Stanford University, 1958.

126. WEYL, HERMANN, "Elementare Theorie der konvexen Polyeder," *Commentarii Helvetici*, 1935, **7**, 290–306. Translated by Harold W. Kuhn: "Elementary Theory of Convex Polyhedra," *Annals of Mathematics Studies*, No. 24, 3–18, 1950.

127. WOLFE, PHILIP, "The Simplex Method for Quadratic Programming," *Econ.*, 1959, **27**, 382–398.

128. ZOUTENDIJK, G., "Maximizing a Function in a Convex Region," *J. Roy. Stat. Soc.*, 1959, **21**, 338–355.

INDEX